"*Adventures in the Aid Trade* is at once me
critique, and globe-trotting picaresque n
and heart, Richard Holloway has produc
scholars and practitioners of development

– Larry Swatuk, Profes ... of Environment,
Enterprise and Development, University of Waterloo, Canada

To David — my
long lost comrade
RHolloway
Richard Holloway
Echenevex
Feb '21

Adventures in the Aid Trade

Adventures in the Aid Trade takes us on a fascinating journey through 40 years of work at the coalface of international development. Drawing on his experiences from long periods in the field, the author reflects on what has worked, what has not and why, and considers how these experiences relate to students and practitioners today.

Looking beyond high-level policy matters and international relations, this book focuses instead on the author's actual experiences in the field and the inspired local people he encountered. The narrative traces how these people, working through their own organisations, make a difference to the lives of their contemporaries, and learn how to generate the income to do it. Chapters draw on the author's experiences of working with local practitioners from 40 countries across sub-Saharan Africa, the Caribbean, South, South East and Central Asia, and the South Pacific. Peppered with lively stories and anecdotes, *Adventures in the Aid Trade* provides valuable lessons from the shifting aid landscape and reflects on where the industry is likely to go next.

Whether you are a current development practitioner or a student just starting out in your understanding of the development and humanitarian sectors, this book provides an invaluable snapshot of the world of civil society organisations, governance and the voluntary sector, and the lived lives of ordinary people in extraordinary times.

Richard Holloway is an international development professional with more than 40 years' experience managing social development projects and programmes in Africa, Asia, the Caribbean and the South Pacific. He has extensive experience of working with non-state and state actors to strengthen processes of citizen–state engagement, and over 20 years' experience of implementing and managing large donor-funded projects (USAID, DFID, UNDP, EU, World Bank). He is currently an independent consultant after many years as a long-term project manager. His notable books are *Beyond NGOs: CSOs with Development Impact, Doing Development: Governments, CSOs and the Rural Poor in Asia* and *Towards Financial Self-Reliance: Handbook on Resource Mobilization for CSOs in the South.*

Adventures in the Aid Trade

Forty Years Practising Development in Forty Countries

Richard Holloway

Routledge
Taylor & Francis Group
LONDON AND NEW YORK

First published 2020
by Routledge
2 Park Square, Milton Park, Abingdon, Oxon OX14 4RN

and by Routledge
52 Vanderbilt Avenue, New York, NY 10017

Routledge is an imprint of the Taylor & Francis Group, an informa business

British Library Cataloguing-in-Publication Data
A catalogue record for this book is available from the British Library

Library of Congress Cataloging-in-Publication Data
Names: Holloway, Richard, 1945- author.
Title: Adventures in the aid trade : forty years practising development in forty countries / Richard Holloway.
Description: Abingdon, Oxon ; New York, NY : Routledge, 2020. |
Includes bibliographical references and index.
Identifiers: LCCN 2019050511 (print) | LCCN 2019050512 (ebook) |
ISBN 9780367434038 (paperback) | ISBN 9780367434045 (hardback) |
ISBN 9781003002963 (ebook)
Subjects: LCSH: Economic assistance–Developing countries. |
Developing countries–Economic conditions.
Classification: LCC HC60 .H66435 2020 (print) |
LCC HC60 (ebook) | DDC 338.9109172/4–dc23
LC record available at https://lccn.loc.gov/2019050511
LC ebook record available at https://lccn.loc.gov/2019050512

ISBN: 978-0-367-43404-5 (hbk)
ISBN: 978-0-367-43403-8 (pbk)
ISBN: 978-1-003-00296-3 (ebk)

Typeset in Bembo
by Swales & Willis, Exeter, Devon, UK

MIX
Paper from responsible sources
FSC www.fsc.org FSC™ C013985

Printed in the United Kingdom
by Henry Ling Limited

Contents

Figures

Acknowledgements

Without a doubt my first thanks go to my wife, Clare, who was, I am sure, glad to see me getting my teeth into something rather than moping around the house glooming about retirement, and was hugely supportive, allowing me to draw on her experience as well as my own – especially for the chapter on Zambia. Secondly, to my sister, Penny, who was so enthusiastic that I feel apprehensive about living up to her idea of what I might produce. Thirdly, to my children, Heron, Guy and Hannah Rose, who have often prodded me to put my stories into print. And finally, to all the other odd bods who I have bored with my stories and who have over time suggested I write it all down.

Once I had made a draft and showed it to people, I have to thank Sarah Archer, development practitioner nonpareil, who helped to annotate it and enthused over it, saying she wished she had had something like this to refer to when she started working.

Those who have refreshed my memory and corrected mistakes are George Awalla of Kenya, Douglas Kammen of Timor-Leste and Singapore, Beryl Knotts of South Sudan and Oxford, Betty Katoro of South Sudan and Leeds, Ross Kidd of Botswana and Canada, Megan Biesele of Botswana, Namibia and Texas, and David Lewis of Bangladesh and the LSE. The final text is my responsibility, obviously, not theirs.

I hope others whose names appear here (or who I have contacted, or who I have quoted) have no violent objections to what I have written – Ezra and Josephine Tsegaye of Ethiopia and Bath, Like Siltana Habte Mariam Workeneh of Ethiopia and USA, Kassa Kebede of Ethiopia and USA, Bill Grisley of Botswana and Illinois, Claire Herford of Dominica and Bristol, Robin Powell of South Sudan and Dulwich, Tom Belknap of South Sudan, Kenya and California, the Grey family of so many places and Sussex, Ken Calvert of Vanuatu, Glen Williams of Java and Oxford, Anton Soejarwo of Indonesia, Paulus Santosa of Indonesia, Janet Durno of Thailand and Canada, retired Surgeon General Suroyo of Indonesia,

Aroma Dutta of Bangladesh, Qazi Faruque Ahmed of Bangladesh, Brenda TambaTamba Liswaniso of Zambia, Isomiddin Sahahuddinov of Tajikistan, Felia Salim of Indonesia, Win Myo Thu of Myanmar, Monique Sternin of the USA, Anna Mugambi of Kenya, David Allen of Myanmar and Bangladesh, and Robert Chambers of Sussex IDS.

I know of some I have mentioned who have passed on – Mabrahtu Yohannes of Ethiopia, Professor Richard Titmuss of the LSE, Greg Moore of Patchwork, Ann Dunham Soetoro of Indonesia, Kansas and Hawaii, Jerry Van Koeverden of Thailand and Canada, Marjorie Shostak of Botswana and Harvard, Sir Fazle Abed of Bangladesh, and Ian Dury of Billericay. There may be others.

Many thanks to all.

Richard Holloway, Echenevex
www.richardholloway.org
rjvholloway@gmail.com

Acronyms and abbreviations

88 Generation	Organisation of ex-political prisoners (Myanmar)
ADAB	Association of Development Agencies in Bangladesh
ADB	Asian Development Bank
AIDS	acquired immune deficiency syndrome
AKDN	Aga Khan Development Network
AKF	Aga Khan Foundation
AKFED	Aga Khan Fund for Economic Development
AKRSP	Aga Khan Rural Support Programme
ALARM	Myanmar environmental CSO
ANSA	Affiliated Networks in Social Accountability
APACE	Australian AT organisation (Solomon Islands)
APM	All Party Mechanism (Nepal)
ARVs	antiretrovirals
AT	appropriate technology
ATI	Appropriate Technology International (no longer exists)
BRAC	Bangladesh Rehabilitation Assistance Committee / Bangladesh Rural Advancement Committee / Building Resources Across Communities (known as BRAC)
CAA	Community Aid Abroad (Australia)
CAF	Charities Aid Foundation (UK)
CAPART	Council for the Advancement of People's Action (India)
CARE	Cooperative for American Remittances to Europe / Cooperative for Assistance and Relief Everywhere (known as CARE)
CBC	Canadian Broadcasting Corporation
CBO	community-based organisation
CEP	Community Empowerment and Local Governance Project (Timor-Leste)
CGI	Consultative Group on Indonesia

CIAA	Commission for the Investigation of the Abuse of Authority (Nepal)
CIVICUS	international federation of NGOs
CPN-M	Communist Party of Nepal – Maoist
CRS	Catholic Relief Services
CSO	civil society organisation
CSP	Civil Society Programme (AKF)
CUSO	Canadian University Service Overseas (known as CUSO)
Danida	Danish International Development Agency
DDRR	Disarmament, Demobilization, Rehabilitation and Reintegration
DFGG	Demand for Good Governance (World Bank)
DFID	Department for International Development (UK)
DRC	Democratic Republic of Congo
ECCDI	Myanmar environmental CSO
ECFWA	Ethiopian Child and Family Welfare Association
ECSP	Enhancing Civil Society Programmes in Timor-Leste
EIA	Environmental Investigation Agency (UK)
FALINTIL	Forças Armadas da Libertação Nacional de Timor-Leste (Armed Forces for the National Liberation of East Timor)
FAQs	frequently asked questions
F-FDTL	Timor-Leste Defence Force
FLEGT	Forest Law Enforcement, Governance and Trade (EU environmental forestry programme)
FRELIMO	Frente de Libertação de Moçambique (Mozambique Liberation Front)
FRETILIN	Frente Revolucionária de Timor-Leste Independente (Revolutionary Front for an Independent East Timor)
GPSA	Global Partnership in Social Accountability
HIV/AIDS	human immunodeficiency virus/acquired immune deficiency syndrome
ICRC	International Committee of the Red Cross
IDP	internally displaced person
IDS	Institute of Development Studies (UK)
IMF	International Monetary Fund
INGO	international non-governmental organisation
INTERFET	International Force East Timor (UN Peacekeeping body)
ITDG	Intermediate Technology Development Group (now Practical Action)

IUCW	International Union for Child Welfare (no longer exists)
KKN	*Korupsi, Kolusi, Nepotisme* (Corruption, Collusion, Nepotism) (Indonesia)
LDO	Local Development Officer (Nepal)
log frame	logical framework analysis (project management tool)
LSE	London School of Economics and Political Science
LSM	*Lembaga Swadaya Masyarakat* (People's Self-help Groups) (Indonesia)
MAF	Mission Aviation Fellowship (Papua New Guinea)
NGO	non-governmental organisation
Norad	Norwegian Agency for Development Cooperation
NTB	*Nusa Tenggara Barat* (Indonesia)
NTT	*Nusa Tenggara Timur* (Indonesia)
NYTC	Ngamiland Youth Training Centre (Botswana)
OCA	Organisational Capacity Assessment
OCAT	Organisational Capacity Assessment Tool
ODA	Overseas Development Administration (UK)
Oxfam	Oxford Committee for Famine Relief (known as Oxfam)
Pact	Private Agencies Collaborating Together (known as Pact)
PBSP	Philippine Business for Social Progress
PECSA	Programme to Enhance Capacity for Social Accountability (Cambodia)
PKK	welfare association of Indonesian government officials' wives
PLWA	person living with AIDS
PNG	Papua New Guinea
PPA	Programme Partnership Arrangement (DFID)
PPIMA	Public Policy, Information and Advocacy (Rwanda)
PRAN	Program for Accountability in Nepal
PRIA	Participatory Research in Asia
PRIP	Private Rural Initiatives Programme (known as PRIP, now the PRIP Trust)
PVO	private voluntary organisation
Sida	Swedish International Development Cooperation Agency
SPECTRUM	Myanmar development CSO
Tiri	Maori word for philanthropic organisation now called "Integrity Action"
UEFA	Union of European Football Associations
UNA	United Nations Association (UK)

UNDP	United Nations Development Programme
UNICEF	United Nations Children's Fund
USAID	United States Agency for International Development
VDC	Village Development Council (Nepal)
VDRC	Vijaya Development Resource Centre (Nepal)
VITA	Volunteers in International Technical Assistance (USA)
VPA	Voluntary Partnership Agreement (part of the EU FLEGT programme)
VSO	Voluntary Service Overseas
WHO	World Health Organization
WPE	women, poor and excluded (World Bank terminology)

Preface

I am in St Lucia, West Indies, in 1979, trying to interest several lads liming[1] on the streets of the town to be self-employed, trying to suggest ways in which they can make money, learn to make something that sells or provide a service that people will want, and still be cool and have the street cred that is so important to them. Taking the example I learned from Father Gerard Pantin of Trinidad,[2] I talk of saving money to stake themselves by cutting down on the weed. It does not go down too well, but they are responsive to someone talking seriously to them about a way out of unemployment. One of the reasons why I am getting only a moderate response is that it is Carnival, and they (and I) are looking forward to the Jump Up.

Three days later, with others, I am pushing the bass pans around the streets. They are four full-size (but empty) 44-gallon drums mounted, with their players, on a wooden wheeled platform. We come thudding out the bass notes at the tail end of a procession of steel drums making their amazing carnival music. The players are crazy men, with no sleep for three days, lots of rum and weed, but it's hard work for four or five of us to keep the platform moving over potholed streets and small hills. Keeping the pan moving, being part of carnival …

Next day in Barbados I returned to the world of my work as Social Development Adviser for the UK's Department for International Development. At the airport a man with an English accent approached me. "I saw you on the streets of Castries yesterday," he said. "Working at the coalface, weren't you?"

I have worked in international development in over 40 countries for over 40 years and have tried to be at the business end of the work – the coalface – as opposed to the advisory or policy level. It has been exhilarating and instructive, but at the end of the day my work has nearly always relied on the aid trade – i.e. foreign assistance, from governments, from businesses and from ordinary citizens – and the aim has always been to make sure such funds were well used. At different times it has been a joyous opportunity to help poor and miserable people rediscover and improve their lives, and at some times and in some places it's been a load of fun.

This book is aimed at actual and would-be development practitioners – fieldworkers and students who are starting out or would like to hear more of the experiences of one long practised in the life they are heading into. I talk of what has worked, what hasn't and why, and what is relevant now.

I am keen to play back my experience for the next generation interested in a range of topics relevant to development – such as:

- civil society organisations and their tussles with governments;
- self-help and volunteers versus government service delivery;
- finding and applying really appropriate technologies;
- positive deviants and how they can solve difficult issues;
- plumbing the depths of corruption and coming out the other side;
- soft fantasies and hard rocks in good governance; and
- building integrity systematically.

There is much more. I offer lessons learned, lessons pondered and valuable takeaways from 40 countries of sub-Saharan Africa, the Caribbean, South, South East and Central Asia, and the South Pacific. Nearly all the experiences come from long periods in the field, alongside local practitioners, and are flavoured with vignettes and anecdotes from my personal life, containing memories of great personal experiences as good as pushing the bass pan platform along the streets of Castries.

Notes

1 "Liming" is Caribbean slang for hanging around the streets, not doing very much of anything.

2 Gerard Pantin started SERVOL (Service Volunteered for All) in Trinidad, working with unemployed youth.

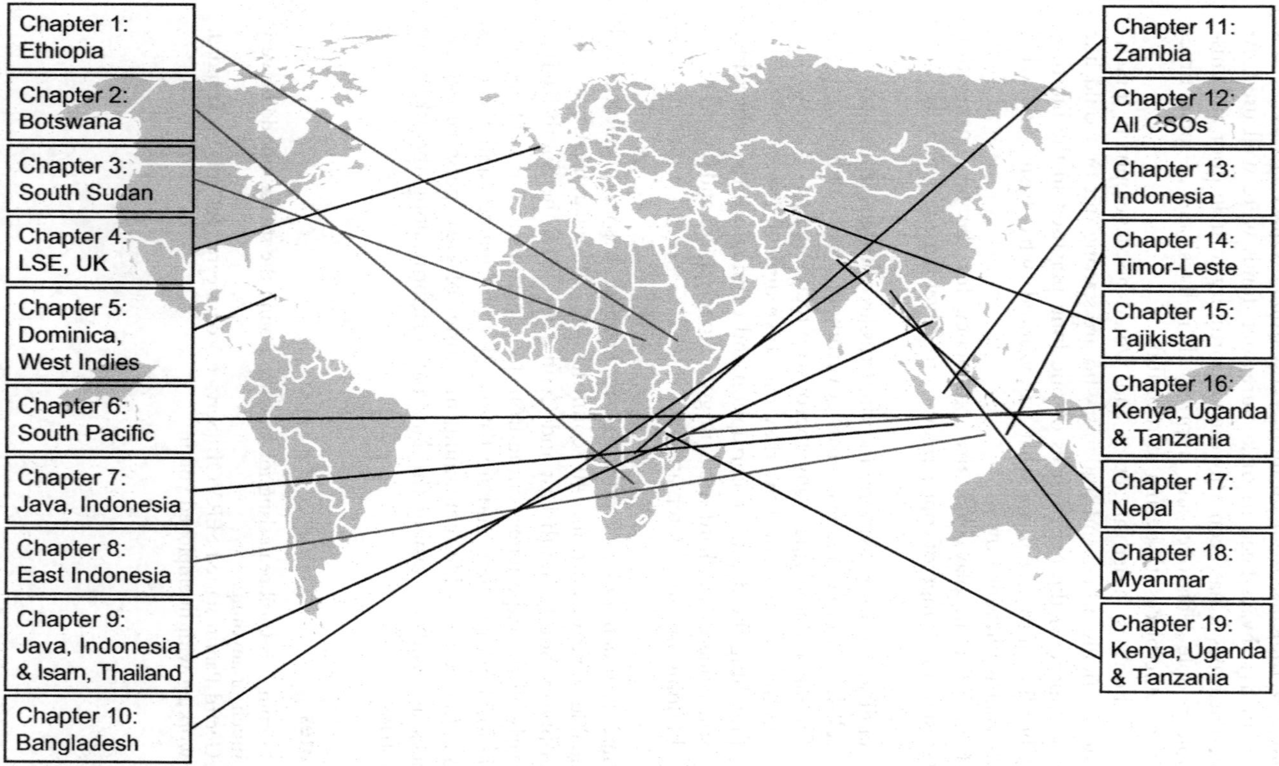

Figure 0.1 Locations of chapter activities

Foreword

Adventures in the Aid Trade stands alone for the extraordinary range of experience and insight it presents. I know of no other book quite like it. Over the past 50 years, from 1966 to 2019, Richard Holloway has worked in at least 40 countries. A few like him who have worked independently as freelance consultants, mainly from the North, could make similar claims, but he differs in having engaged for a matter of years living and working in several of them – six years in Bangladesh, five in Indonesia, five in Zambia, five in Malaysia and shorter but substantial periods in Nepal, Botswana, East Timor and elsewhere. And in Indonesia he became fluent in the national language.

The life experiences recounted and reflected on here have not been in the mainstream of any organisation. This has given him flexibility but also insecurity. It shines through that he has avoided the freelancer's trap of telling sponsors what it is known they want to hear. He has taken the opportunities offered by a range of funders of programmes in which he could live, work and contribute substantially without being a career employee. It is much to their credit that his funders have given him space to be transparently and revealingly honest. He does not hide things but lays them out. He is self-critical and consequently credible, adding so much to the value of this book and what we can learn from it.

For those from countries which are the subjects of chapters, and for those from other countries, there is here fascinating detail about the reality of conditions, cultures and relationships at various times in the past half century. It will endure as a valuable historical document presenting an outsider's view, rounded and given depth by including his own country, the UK, where he worked with a community organisation inspired by squatters.

This book is then far from being the superficial descriptions and reflections of the brief visitor. It is a rich harvest of the experience of someone who has been critically aware of the cognitive traps and biases to which development professionals are so widely vulnerable, and who has consistently and effectively striven to offset and avoid them.

In the course of his varied career, he has spent periods living and working in depth, sometimes in remote areas, and learning with and about poor and marginalised people. So these adventures are grounded in a remarkable and unique life experience, from which I and all other development professionals, and perhaps especially those setting out on hoped-for careers in development, have much to learn.

This learning is made easier by Holloway's transparent honesty. This is a book of warts and all. He sets an example by laying bare what he did and what did not work as well as what did. He is reflective about his own life, identifying what has helped him towards doing and learning better. Of his time at LSE he writes: "Having more field experience made my LSE experience more pertinent and valuable to me." There is a lesson there both for those who convene and for those who choose to take part in courses related to development.

There is much to learn also from the topics and themes that recur. Of these, let me single out the most prominent and significant. This is corruption. It stands out in most countries and many contexts. It is often pervasive in whole organisations and administrative systems whether in Africa or Asia. Holloway recounts resolute attempts to overcome it, and its maddening resilience. It is not all gloom and doom, though. There are successes and examples of what can work to diminish corruption and other malpractices. He works with advocacy organisations and programmes of social accountability to enable people to claim their rights. He is involved in a repertoire of approaches which can be mobilised, such as role plays, comic strips and cartoons. These are examples of the many practical experience-based ideas with which the book is replete and from which there is much to learn.

Perhaps most useful of all, and presented in a succinct and accessible form, are the concluding sections in almost every chapter entitled "What did I learn from the coalface?" Here is well-grounded advice which gives scope for much reflection, internalisation and action. Taken together, these sections present a wise and informed source, to influence one's choices and decisions. These short sections are a treasure trove for all of us from all countries, North and South, in the spirit of universalism of the SDGs, whether we are freelance or work in governments, CSOs, the private sector, the media, academia or specialised professions, or whether we are freelance. It is for all of us who are eager to learn practical lessons and to improve our practice.

Looking to the future, Holloway starts with the present. He notes the increasing commercialisation of aid, and the top-down accountability and the market language that go with this and with the large consultancy firms that receive big contracts. The phrase "the aid trade" represents not so much cynicism as realism about currently prevailing aid transactions and relationships. Throughout the book his optimism and positive attitudes and approach shine through. It is the same with the future. He recognises that we are in a new space. The future will not be like the past. We live in a new, enthralling post-postcolonial world of peers, mutual accountability and co-learning for which he was a pioneer ahead of his time.

For all who work in development or aspire to do so, *Adventures* is a grounded and invaluable source of learning and inspiration. It is a rich source of ideas for how we can do better. I commend it to all development professionals, whatever their roles, as an engaging read and a fertile source of learning.

Robert Chambers

Introduction

This book rarely concerns itself with high-level policy matters, or with international relations, but with the actual experiences at the coalface where inspired local people, working through their own organisations, make a difference to the lives of their contemporaries and learn how they can generate the income to do it. The book provides valuable lessons, brings them up to date, and tells good stories along the way.

The book contains snapshots of the kinds of development practice with which I have been involved, together with some hindsight on how I could have done it better, what has been the impact 40 years later, and, finally, some references for you to follow up on any of these ideas that still seem to you to make sense. To some it may seem ancient history, but as I have written and reworked the text, I am clearly able to say that many of the ideas are still very valuable and could be very useful to you, if you are likely to be working in international development.

Let me clarify what you will get from reading this book. Of all the 40 plus countries[1] and 40 plus projects where I have worked, I have selected those that, in my opinion, are the most interesting and the most relevant to people working in development. As I have said before, however, the experiences are of involvement at the grassroots, not high-level policy advice or interventions. From time to time the grassroots experience provides many suggestions for policy-level work, but I have not been personally involved in this.

In each of the following 19 chapters I tell the story of a particular project and end each chapter with three observations:

1. What did I learn from the coalface that I can take with me for future work in development?
2. What happened to it all? Where is that experience now?

3. What were we thinking at the time? This section lists the books, videos and films that inspired me and my colleagues at the time, as well as some that dealt with the same topic subsequently.

Where they seem appropriate, I also add some personal recollections and anecdotes to spice up the narrative.

The reader deserves to hear more about who I am and what brought me into these experiences in order to set the scene.

I was an army brat (actually a NAAFI brat, for those who know what that is[2]) and therefore grew up on army bases in foreign lands – but my father was definitely not a soldier. I went to a Methodist secondary school in Canterbury, then to Oxford, where I read English Language and Literature, off for three years to Ethiopia (see Chapter 1), then back to LSE (London School of Economics) to study Social Administration (Chapter 4), where my course was interrupted by two years in Botswana (Chapter 2).

It is important to understand that I never worked for a career organisation – I got jobs based on adverts, personal contacts and being in the right place at the right time, and I was very dependent on the reputation of and references from my last employer. Usually jobs were for two to five years and this process continued until 2014, when I found it difficult to get long-term work (principally because I was too old, I think) and took up short-term consultancies, which I found much less satisfactory. You were never around long enough to really understand the country or the project, and you never found out whether your short-term consultancy had actually been helpful.

In all of this work, spanning 40 odd years, there were four aspects that need to be considered deeply by someone joining this field:

- always being an expatriate in someone else's country;
- being a cross-fertiliser between developing countries, not an adviser from a developed country;
- working for governments or NGOs; and
- looking after a family, which involves a partner and children, periods of unemployment and pensions.

Being an expatriate

You could be useful to a country, and the organisations working in that country, to the extent that you brought something needed and wanted to the country. At the start of my overseas work, depending on the country, there were relatively few local graduates who could do what I was asked to do. As time went on, local people acquired the relevant education and experience and there was less need for us – particularly since we probably did not have the language, as they did (see the section on language in Chapter 20). Expatriate, of course, means a foreigner, and this can cover much more than the "development set"[3] – it also includes members of the business world and artisans (frequently seen as China exports its own people to projects worldwide), as well as informal traders. The word is, however, usually used for people in the development business and can cause bad feelings in the receiving countries because such people are suspected of taking jobs away from locals. It is now much more common for foreigners to come in as short-term consultants to advise locals who have the permanent positions. Any development expatriate needs to be very sure to ask themselves if they are working themselves out of a job and helping a local to take on their role.

One subset of being an expatriate is being a volunteer, working for a volunteer sending organisation – e.g. VSO, CUSO, AVA (Australian Voluntary Associates), DED (the German Development Service) or UNV (United Nations Volunteers). In this case you are paid a local salary and live with local people, and are not usually accused of taking away their jobs. You risk, however, being accused of getting more out of the placement than the country does.

The last point is that you are not from that country, you have no right to vote there, and if you make yourself unpopular (even if it is for a very good cause) you can be kicked out – or, as it is frequently called, made "persona non grata".

Because the aid trade is so large and so extensive, it is perfectly possible to seek permanent employment within the structure of a donor organisation. Such jobs fluctuate with the funds available for development assistance and the ability of a development NGO to win contracts from donors. I did not do this, but it is a feasible career path.

Cross-fertilising between developing countries

As you will see from the list of countries in which I have worked (note 1 in this introduction), I have never worked in the UK, although I was exposed to social institutions in the UK while at LSE (see Chapter 4). I have never felt that being British means that I have something useful to offer to developing countries. In fact, my value comes from the wealth of experiences that I have had in comparable countries beyond the UK. I am useful to small Caribbean Islands because of my experience in small Pacific Islands (and vice versa – Chapters 5 and 6);

I am useful to countries coming out of the experience of centralised one-party states because I have worked in comparator countries (Chapter 15); I am useful to NGOs and CSOs everywhere (and their struggles with governments) because I learned my trade in Bangladesh, home of the best NGOs in the world (Chapter 10); I am useful to all CSOs trying to become financially sustainable because Pact and the Aga Khan Foundation (AKF) gave me the freedom to research and develop these ideas (Chapter 12); I am useful to all countries working on issues of corruption because the Partnership in Indonesia and the AKF allowed me to work on these topics and learn what makes corruption happen (Chapter 13).

Working for governments or NGOs

I have worked only once for governments – the British and Dominican Governments in the Caribbean (see Chapter 5) – and in short-term consultancies for the USA. In all other places I have happily worked for NGOs and CSOs. Compared with the enthusiasm, drive and commitment of NGOs and CSOs, government ministries and departments are, in my experience, and with a few exceptions, dull, boring, bureaucratic places, often riven with corruption. When colleagues point out that working with government allows for large and national-level change whereas NGOs are small and lacking influence, my response has been that NGOs and CSOs can show what can be done and can demonstrate models to governments, and that they are, for the most part, organisations of integrity.

Looking after a family

I was single during Chapters 1–6, married to a partner during Chapters 7–20, and had kids during Chapters 9–20. The biggest danger is having a partner who does not have a worthwhile job that interests him or her, and having children whom you cannot afford to bring up and educate as you want. My partner had a parallel and very interesting career in development, working for local NGOs and then UNICEF until the period of Chapter 15; she then retired with a very useful UNICEF pension.[4] We paid for our kids to go to local primary schools, and then UNICEF, USAID and UNDP at different times paid for our children's secondary schooling and university placements, for which I am eternally thankful. Because of my serial jobs, I never got a worthwhile pension. Learn from me, brothers and sisters – if you want to be a development practitioner and have a family, there are some tough hurdles to overcome.

Morality, ethics and codes of conduct

When I showed drafts of this book to my peers, they asked why I had not addressed the issues of shameful behaviour by expatriate aid workers (notably Oxfam in Haiti and UN Peacekeepers all over the world). In my experience, such events usually occur at times of disaster when the pressures and tensions are very high and expatriate staff are "parachuted" in. I had not had that experience, although I came in on the tail end of such a situation in Timor-Leste in 2004 (see Chapter 14 – particularly the references to the UN's "camp followers"). Organisations employing expatriates should have due diligence structures in place and supervise them well, they should have intensive briefings with potential employees on what is and is not deemed acceptable, and follow-up should be implemented if the codes of conduct are abused – and, in my experience, this is not done sufficiently. Workers in the aid trade are by no means saints; they need to be constrained by clear rules and regulations. It is also true, as I have been told by my daughter, working in South Sudan and Somalia, that aid workers, depending on the danger in their placements, are frequently housed in protected compounds and can only leave these compounds accompanied by security officials.

What is in the book?

- **Social welfare and community development**: This is covered in Chapter 1 on Addis Ababa, Ethiopia; Chapter 2 on Maun, Botswana; Chapter 3 on South Sudan; and Chapter 5 on Dominica, West Indies. In most of these chapters I have had a hands-on role, working alongside the local people who are, hopefully, the beneficiaries of the work I am undertaking.
- **Appropriate technology**: This is covered in Chapter 6 on the South Pacific; Chapter 7 on Java, Indonesia; Chapter 8 on eastern Indonesia; and Chapter 9 on Java, Indonesia and Isarn, Thailand. Here, I am gradually growing away from the coalface myself, but managing others' efforts to work there.
- **The role of NGOs and CSOs, particularly on financial self-reliance**: This is covered by Chapter 10 on Bangladesh; Chapter 11 on Zambia; Chapter 12 on CSOs everywhere; Chapter 14 on East Timor; and Chapter 15 on Tajikistan. My role here has been largely as a trainer.
- **Corruption and social accountability**: This is covered by Chapter 13 on Indonesia; Chapters 16 and 19 on Kenya, Tanzania and

Uganda; Chapter 17 on Nepal; and Chapter 18 on Myanmar. My role here has been as project manager, trainer and adviser

- **Disaster relief and rehabilitation**: There are contributions to these topics in Chapter 3 on the aftermath of civil war in South Sudan; in Chapter 8 on Indonesia's invasion of East Timor; on Bangladesh's floods and cyclones in Chapter 10; and in Chapter 14 on the rehabilitation of Timor-Leste.
- **Training and preparing for the work**: Chapter 4 talks about the training and orientation I got from LSE and being part of Patchwork.
- **Reflections on how the aid business has changed over time**: The conclusion (Chapter 20) talks about how the focus of development agencies has changed over time, how the language and spin of aid has modulated, and my thoughts on training and preparation for overseas development work.

Useful comparable literature

Alexander, Jessica. 2013. *Chasing Chaos: My decade in and out of humanitarian aid*. Broadway Books, New York, USA.

Buxton, Charles. 2019. *Ragged Trousered NGOs: Development work under neoliberalism*. Routledge, Abingdon, UK and New York, USA.

Eyben, Rosalind. 2014. *International Aid and the Making of a Better World: Reflexive practice*. Routledge, Abingdon, UK and New York, USA.

Mosse, David (ed.). 2011. *Adventures in Aidland: The anthropology of professionals in international development*. Berghahn Books, New York, USA and Oxford, UK.

Postlewait, Heidi, Cain, Kenneth and Thomson, Andrew. 2004. *Emergency Sex (and Other Desperate Measures): True stories from a war zone*. Miramax Books, New York, USA.

Key terms

The world of the aid trade, particularly the non-government part of it, is replete with its own language and terms – many of which are incomprehensible to outsiders.

Here are the chief terms that I have used:

- **NGO (non-governmental organisation – INGOs are international NGOs)**: This refers to the larger national or international non-government development organisations. It also assumes that they are not-for-profit organisations.

- **CSO (civil society organisations)**: This refers to the smaller and more local versions of the above, clarifying that they are civil (i.e. not government and not military).
- **CBO (community-based organisations)**: These are even smaller local organisations, linked to a particular community.
- **PVO (private voluntary organisations)**: This is an American term for NGO, reflecting the US tax code.
- **Non-state actors**: This is a European Union term for NGOs designed to include trade unions and professional organisations.
- **The third sector**: This means the civil society sector, not the government (first sector) or business (second sector). It can include non-developmental interests like sport or culture.

During the periods covered by this book, all of these terms were in the process of being defined, refined and tried out in practice.

- **Bilateral development agencies**: These are the national aid agencies of a country (e.g. DFID for the UK, USAID for the USA, Norad for Norway, Danida for Denmark, AusAID for Australia).
- **Multilateral development agencies**: These are the UN agencies (e.g. UNICEF, UNDP, WHO) and finance agencies (e.g. World Bank, IMF).

Notes

1 USA, Canada, Guyana, Portugal, Bulgaria, Hungary, Afghanistan, Tajikistan, Kyrgyzstan, Pakistan, India, Sri Lanka, Egypt, South Sudan, Ethiopia, Kenya, Uganda, Tanzania, Rwanda, Zambia, Botswana, Lesotho, Swaziland, Mauritius, Papua New Guinea, Solomon Islands, Tonga, Samoa, Vanuatu, Fiji, Timor-Leste, Indonesia, Malaysia, Myanmar, Bangladesh, Philippines, Thailand, Syria, Mozambique, South Africa, Mali, Dominica, St Lucia, St Vincent, Turks and Caicos Islands.

2 NAAFI is "Navy, Army and Air Force Institutes", which is the organisation for recreational, catering and grocery institutions for the services and their families. It is a separate and independent company, which, however, is always connected to the Armed Forces.

3 "The Development Set" is a poem by Ross Coggins to describe expatriate development workers, written in 1976 when he worked for USAID in Indonesia. It is satirical and biting, and can be accessed on Google.

4 My wife, however, did not start her career in the development aid business, but in theatre management (see note 5 in Chapter 4).

1 Addis Ababa, Ethiopia and street children, 1966–69

Trying hard to keep a welfare institution going

… In which an Ethiopian organisation helps juvenile delinquents and street children into schools, and, in particular, tries to fund itself, including by running a potato chip factory.

Addis Ababa in 1966. And me going out there fresh from an English Language and Literature degree at the University of Oxford to teach English in a secondary Commercial School to Ethiopian students keen to become bureaucrats and businessmen.[1] After a short and intense course on how to teach – arranged by my sending organisation – I faced my first class, and aided by their competent textbooks, I fed English language into their eager ears, eyes and mouths. It was logical and sensible work – they wanted to learn English, I was English, and with a bit of guidance I could teach them English with the expectation that this would help them get on in the modern world they wanted to inhabit. The students were very varied, ranging from poor kids who represented their parents' desperate investment to help their children (and them) move up in the world to spoilt brats of rich families who were coasting through to a life of privilege.

Looking out of the school compound, however, and watching urban Addis Ababa life as it passed by, I was struck by the huge difference between the students I taught and their cousins who lived on the streets, shining shoes, carrying loads, portering, washing motorcycles, petty thieving and huddling up with their peers in shop doorways and on balconies to sleep in the cold Ethiopian nights. I started asking around to find out who was interesting themselves in their situation and soon found two organisations linked by a charismatic Ethiopian, Mabrahtu Yohannes. He was, as a government employee, the Director of the Juvenile Prison, tasked with administering custodial sentences for youngsters who had been caught by the police and sentenced by the courts. He was also, as a caring citizen, the lead figure in a recently formed charity called the Ethiopian Child and

Family Welfare Association (ECFWA), which was formed to try to get to the street kids and address their problems before they broke the law, got arrested, and were sentenced to the juvenile prison. It was an Ethiopian association and the president was Like Siltana Habte Marian Workeneh, an archbishop of the Ethiopian Orthodox Church.

I became very friendly with Mabrahtu and spent a lot of time at the juvenile prison, learning Amharic and finding out about juvenile delinquency in Addis Ababa. After six months of teaching English to students at the Commercial School, my heart was not in it any more, and I applied to my sending organisation, which was locally managed by the British Council, to change jobs and work for the Association. The Commercial School had thought that they had me for two years and so were disappointed, but they could handle it. More important was the fact that the ECFWA had no funds to pay even a volunteer's salary, and so I needed to fundraise fast.

Like many Ethiopian charitable organisations, it requested donations, particularly regular bank transfers, but it also had a small shop selling second-hand goods, mainly clothes – very much on the lines of an Oxfam shop. This was good – it was a going concern, but it couldn't cover its costs, which basically covered a hostel for 40 young boys, let alone a living allowance for me. It also depended on regular gifts from Ethiopian and foreign well-wishers of a quality that would attract punters.

The purpose of the ECFWA was to address the situation of young boys who had run away from home, or been thrown out of their homes, or been orphaned in the country around Addis Ababa. They had come to town in order to seek education, which they could not find near their homes. They came to town, tried to get into school, and soon found that if they attended school, they could not feed themselves because they had to be on the streets hustling for income. If they developed a way of earning an income, like shoe shining, they could not take time off to go to school. Many of them, therefore, became street vagrants, in spite of themselves – they wanted hugely to go to school. The ECFWA offered a place to sleep, gave them food, and made sure they got into school. Proficiency at school was not a problem, once they had a place. These kids were very motivated and did very well. And once they were in school, very few returned to the streets. We took only boys – there were few street girls in the same state, but we were all, in hindsight, less alive to the problems of street girls than we should have been.

I contacted Oxfam through the Association and persuaded them to pay the basic costs for a limited number of young boys for a limited period of time, as well as an allowance for me.

I was apprenticed to Mabrahtu and put in charge of our first hostel in Lideta, a very poor area of Addis Ababa. The Executive Committee of the ECFWA agreed to take me on, with the strong instructions to raise money for the organisation for when the Oxfam money ceased.

And then Mabrahtu informed me that he had been offered a scholarship to study Criminology in Canada for two years, and asked would I be ready to become the manager of the Association for the next two years until he returned. Many warned me that he would leave, become a Canadian and never come back, but I never doubted him, and indeed he returned as promised and took the organisation back from me two years later.

But in the meantime, under the Ethiopian Executive Committee, I was in charge of 40 boys' upkeep, with little previous experience, and insufficient Amharic to work successfully on the streets. I needed help. Ezra Tsegaye appeared, son of a successful Ethiopian architect, just finished at an international school and very motivated to make the Association work.

With his help we established a routine for running the hostels – now grown to two hostels housing 90 boys in two parts of town – but it all depended on making sure the money came in. We had to move from being an organisation based on charitable donations and charitable motivations into something that could grow, be professional in its approach to the children in its care, and be sustainable.

First, we kept costs to a minimum – the boys ran the hostels, bought the food and organised their living conditions, with recourse to Ezra or me in times of trouble, which had to be kept to a minimum. There was a cook/housemother in each hostel. On our Executive Committee were Ethiopia's only probation officers, lecturers from the University's School of Social Work and a volunteer treasurer from UNDP. The only other member of staff was the guy who managed the second-hand goods shop.

How to make money for the Association? With the very low foundation of the Oxfam money, we lucked into opportunities (like the donation of puppets) and used our initiative to find income in the following ways.

Selling imported second-hand clothing that could not be sold!

Catholic Relief Services (USA) imported huge amounts of second-hand clothing with very strict instructions that these were to be given out through Catholic Church channels and *not sold*. I started the long negotiations to try to turn around this important prohibition.

The argument was about giving away often bizarre and unsuitable clothing from the USA to very poor Ethiopians versus giving these same clothes to our shop, where middle-class Ethiopians would buy snazzy fashions (albeit a bit dated) and where the clothes could be sold for money to run an institution that helped poor street kids get an education. We won the argument largely based on our reputation for professional services to street kids, and started receiving enormous bales of extraordinary clothing, packed under pressure. You cut the baling straps and clothing exploded into whatever place we had borrowed for the exercise (often friends' lodgings). The Thrift Shop was well stocked and became a solid source of income.

Making playgrounds

At the weekends when the kids were not at school, we advertised that the hostels could build playgrounds from the ubiquitous eucalyptus poles that were the scaffolding medium for the building trade. We advertised to embassies and UN agencies that we could do this, and built swings, climbing frames and seesaws. No one else seemed to be doing this, and rich foreigners were prepared to pay for the service. No one got hurt, and we publicised our work. I had never made playground equipment before, but it was not that difficult.

Puppet shows and satirical sketches

As we became more famous I was approached by a German pastor with a beautiful set of glove puppets that he was no longer using, and he asked if they were helpful. This started another income stream as an American friend, Jeff, and I put on puppet shows in English and Amharic (you don't need a lot of language for a puppet show!) for institutions and for foreigners' children's parties. Next came satirical sketches (this *was* the late Sixties!). With the same American friend, we put on "Sophisticated Satirical Evenings" at upper-class and intellectual Ethiopian and foreign parties. At least this was something that Jeff and I had some experience in from our university days. The problems were those of political sensitivity – here were two young foreigners satirising facets of Ethiopian life – so we took plenty of advice and warning. Another income stream.

Kesh-Kesh

All the income streams above were fun and gave us good publicity, but did not bring in the badly needed amounts of money. We had to have something a bit more productive. Then, one day, an Ethiopian businessman called Dereje, who was about to emigrate to the USA, having heard of our work, told us he would like to donate to us his potato crisp business, called onomatopoeically Kesh-Kesh, so that we would have a commercial concern to underpin our social work. This sounded marvellous and we grasped it with both hands, only to find it a bit more than we could handle. It was a registered company with a product line and 11 trained staff (but no manager – this was the Ethiopian businessman who was leaving). I had no idea how to run a business, but it sounded like such a good idea we could not refuse. None of the donor organisations interested in our social work had any expertise in business either, and so they could not help. Making potato crisps requires big and good-quality potatoes; it also requires some expertise in product health and safety requirements and, above all, marketing. Small packets of potato crisps are snacks and go well with drinking beer; beer in Addis Ababa at that time was sold in *tej-bet*, akin to South African shebeens. Large packets were sold in middle-class grocery stores in competition with imported varieties that had passed all sorts of phytosanitary regulations.

It was an object lesson in sticking to your mission: three months after we had taken over the running of Kesh-Kesh, you would have found me at six in the morning in the wholesale potato market buying big potatoes in order to make potato crisps to sell to make money to pay for hostels for street kids. I was meant to be managing the hostels! We thought of giving the small packets to our own kids to sell in the shebeens, then decided that was almost the polar opposite of what we would like the kids to do. We were confused – we were also closed down by a prima-donna Frenchman who headed the Pasteur Institute in Addis Ababa and who had considerable influence over the municipality, who said that our cleanliness was below acceptable standards. We pointed out that Addis Ababa tap water was below acceptable standards. So, as a social worker perforce evolving into a potato crisp producer, I retreated to writing proposals to foreign donors, handing over some optimistic prospects to Mabrahtu when he came back from Canada.

The Ministry of Finance

Many people had told us that our future must be through a grant-in-aid from the Ethiopian government. Relying on foreign aid was neither realistic nor admirable – the Ethiopian people and State should look after their own social casualties. The president of our Association's board, the Archbishop Habte Mariam Workeneh, felt strongly about this and had been manoeuvring to get state funding for the Association for some time but had not shared these ideas with me – perhaps with others, but not with me. To my great surprise, therefore, shortly before I left Ethiopia in 1969, I was gathered up by the Archbishop in his priestly garments and taken to the Ministry of Finance, where a government functionary was instructed to give me a whole pile of money. This was no bank transaction, just a pile of money shoved into a bag. The Archbishop had finally persuaded the government to pay for something, but I'm afraid it was a one-time donation, not sustainable or annual grant-in-aid.

What did I learn from the coalface?

Keep your eyes on the prize: this inspiring slogan from civil rights work in the USA must be your motivation. If you have decided that you are doing good work, and you can persuade others you are doing good work, then think of all the ways you can be financially self-reliant to allow your good work to continue.

Start from where you are, and your own experience, but be very open to opportunities that may arrive. Have a good time, but be conscious of expanding and building your reputation and your fame so that opportunities are presented to you.

Stick to your mission – and is it very easy to be seduced away from it by ideas that are attractive but too complex for you to handle. Find and take good advice, if you can possibly find it. Be aware that you may be a maverick who does not fit established patterns.

Donor agencies, by and large, only understand giving funds: planning for financial sustainability involves expertise in running enterprises, and few donor agencies can be helpful to you there. There is another part of the aid trade called small business development, but it is very rare for this expertise to cross into other sectors, like street children. A singular lacuna in the aid trade.

What happened to it all?

Three years after I left Ethiopia there was a revolution, usually called the Derg. It included a terrible period called the Red Terror, in which many people were caught in the crossfire of street shooting. And these included many of our boys. In the Communist period, all non-state institutions were closed down, including the ECFWA.

Mabrahtu was a Tigre and became very involved in the Tigray People's Liberation Front, which took over the country in 1991. After helping the new government get established, he became a private lawyer but kept in touch with and was very supportive of the hostel kids who survived, as were members of the board. One of the most famous of the alumni is Tadesse Mesfin, a renowned artist in contemporary Ethiopia.

Ezra married his English teacher and emigrated to the UK, working there as a gardener, and often revisiting Ethiopia.

Kesh-Kesh was sold to a bakery and no longer markets crisps by that name.

Street children still abound in Addis Ababa, but, as far as I can see, they are now more professional and committed to their chosen trade – they are not so ready to leave that life to try to attend school.

On a more personal note, against the background of this work I was able to enjoy astonishing opportunities for exploration and travelling in Ethiopia.

I was able to walk from Tendaho cotton plantation in the Danakil area to the lake that is the end of the Awash River as it reaches Djibouti. The slogan of the Awash Bank in Ethiopia's is "Our money stays in Ethiopia, just like the water of the Awash River". I started this journey on camel back, but soon found that riding a camel was very painful on the testicles as you rocked back and forth, and so got off, put the luggage on the camel, and walked beside it.

In the Simyen mountains, I tried to climb Ras Dashen, the highest mountain in Ethiopia – the Italians claimed that it was over 15,000 feet for their colonial pride, but this was disputed. I never made it to the top, becoming overpowered with altitude sickness – and also losing my way – and sleeping out on the mountain. I was rescued by the indomitable Irish adventurer Dervla Murphy, who was coming through the same mountain range as part of her journey written up as *In Ethiopia with a Mule* – see Chapter 4 in that book.[2]

Most of Ethiopia is a high tableland, but where that tableland comes close to South Sudan, it drops down quite dramatically – from Dembidolo to Gambela. While expiating a broken love affair, I walked for two days down to the Sudan, losing height dramatically, passing colobus monkeys in the trees, looking like jockeys in their black and white silks, and meeting increasingly black people as the Amhara gave way to Anuak. Once down on the Sudanese plain, I borrowed a bicycle to explore, only to find that the renowned tough Chinese bicycle started to break up, and I walked back to Gambela with it over my shoulder.

Readers may remember the extraordinary airlift of Ethiopian Jews to Israel, at the end of the Derg period. These folk were called

Falasha, and were a distinct group in Ethiopia who were given the chance to be taken to Israel. On a subsequent visit to Ethiopia, I enquired about the history and whereabouts of the kids and the staff of the hostels. Mama Etenesh, who used to be the cook and house-mother in one of the hostels, had, I was told, gone to Israel in the airlift. "But she wasn't a Falasha," I protested. "You knew that, and we knew that," they said, "but the Israelis didn't."

Working with street kids and publicising their situation was slightly tricky in Ethiopia in the late Sixties. Proud Ethiopians might think that your chief reason was to do down the reputation of the country. We managed a large photo documentary of kids sleeping on the streets, begging, etc., but handled it through a large and famous secondary school that took on the project as civic education for their students and the parents of their students.

Many people know of the Rastafarians from Jamaica. I had a strange link to them. It is common knowledge that the Emperor Haile Selassie's title was Ras Tafari, and this was picked up by "Back to Africa" Rastafarians in Jamaica. Haile Selassie had given land to some Rastas in a town called Shashamane, and I came across them in a bizarre circumstance. British eccentric General Orde Wingate helped Haile Selassie reoccupy Ethiopia from the Italians, and a school was started in his name outside Addis Ababa called the Wingate School and modelled on a British public school. In this school, they instituted cricket (how eccentric can you get?!). Once a year there was a public cricket day and all the folks from the Commonwealth who knew about cricket charged in – Indians, Pakistanis, Sri Lankans, Australians, West Indians, Brits. I joined as well and was approached by a young black lad who, not surprisingly, I thought was Ethiopian. "Please can I play too?" he said. "I know cricket." It turned out that he was the child of a Jamaican Rasta immigrant who had heard there was some cricket going and could not keep away. He was brilliant at bowling, batting, catching – and this was my first exposure to the unique accent of Jamaica.

I also had an occasion once to go horse riding with the daughter of the Minister of Finance. She was a very international woman whose brother, Dereje, had given us the Kesh-Kesh factory, and she had access to horses. She was as beautiful as most Ethiopian women and ran a clothing boutique in downtown Addis Ababa. My intentions were entirely honourable. As we rode along I noticed a man on a very active horse consistently riding with us, but about 100 yards away. "Is that man following us?" I asked. "He is a bodyguard my father has sent to protect me," she replied. I gripped my horse more firmly.

What were we thinking at the time?

As I am sure is obvious from the contents of this chapter, I was winging it in Addis Ababa, responding to opportunities as they arose. I do not remember any research on street kids in Addis Ababa, but I was aware of the work of the Starehe Boys Centre in Nairobi, which I visited and learned from.

As I have noted, street kids are still a large problem in Addis Ababa. In this period, street kids – e.g. shoe shine boys – saw their work as something they had to do while they tried to get into school, but now shoe shine boys are older and more established – they are, if you like, more professional.

If you want to learn more about the contemporary situation, please see:

Mengesha, Mekonnen. 2012. *Street Children in Addis Ababa: Exploring the policy framework for interventions*. LAP Lambert Academic Publishing, Saarbrücken, Germany.

Roberts, Angela Raven. 1991. *Notes from the Field – Strategies for Street Children in Addis Ababa: Defining the issues*. Michigan University Press, Ann Arbor MI, USA.

If it were undertaken these days, the Kesh-Kesh adventure would be called a social enterprise. Back in the Sixties it was pure opportunism, and I was ignorant of how to run a business, although I had some help from the Peace Corps. Someone thinking of the same kind of intervention these days would be advised to look at:

Caldwell, AmyAnn. 2014. *What is a social enterprise? A simple definition and 3 examples*. The Good Trade, Los Angeles CA, USA. Available at www.thegoodtrade.com/features/what-is-a-social-enterprise.

Ethiopia hosted a world conference on social enterprises in 2019, and Social Enterprise World Forums (SEWF) showcases many of the Ethiopian examples (www.sewf2019.org).

Notes

1 The sending organisation, the United Nations Association, working through the British Council, asked for preferred countries. I replied that I wanted a country that had not been part of the British Empire and I was offered Ethiopia.

2 Murphy, Dervla. 2003. *In Ethiopia with a Mule*. John Murray Publishers, London, UK.

2 Maun, Botswana, 1970–72

Making technical education pay for itself

... In which the Batawana, the people of remote North-West Botswana, started and ran a vocational training school to complement an academic secondary school, and tried to make it financially self-reliant.

The work of the ECFWA in Ethiopia was very urban. I was therefore delighted when the next job that was offered me by the United Nations Association[1] was the complete opposite – working in the remote rural area of North-West Botswana. Botswana was recently independent, it had just discovered diamonds, and it knew with confidence that it was going to be very different from a South African Bantustan.[2] Botswana taught me about rural life, communal landownership and tribal administration.

The North-West District of Botswana, of which Maun is the capital, is very different from the rest of the country because of the singular geographical feature of the Okavango Swamps. The Okavango River comes in from Angola, across the Caprivi Strip, meets the Kalahari Desert and spreads out into a shallow body of water of 600 square miles. The landscape in the swamps is of small islands of beautiful vegetation in lagoons of clear water. Outside the influence of the swamps, the North-West is similar to the rest of Botswana – the dry Kalahari Desert.[3]

The North-West[4] has always been very rich in wildlife and cattle (although the cattle were susceptible to and affected by tsetse fly): now it is a place of safaris and game viewing, with reserved areas for licensed shooting of game at very high prices. The Okavango floods every year, however, and there is the possibility of arable farming on the receding floodplain – something that no other part of Botswana can boast of.

In the Seventies, Botswana had recently become independent and had looked pessimistically to its future as a country whose resources were restricted to goats and cattle grazing in the Kalahari with possible sales of livestock to the countries of the Gulf. At the time of independence,

Botswana had a very limited number of its own graduates, and the upper levels of the civil service were largely British or South African (and South Africa was still under apartheid). Then diamonds were discovered and suddenly the country realised that it would be rich and could have all the attributes of a modern welfare state.[5] This had two results: first, the expansion of a colonial model of an African country – which could pay for itself: and second, the opportunity to do things differently from such a colonial model – encouraged by left-leaning foreign advisers and the new intelligentsia of Botswana graduates.

Thus, in the field of education, many primary schools were built, along with a few traditional colonial secondary schools with an academic curriculum. As in many other developing countries, this model produced large numbers of primary school graduates who had no hope of getting the few secondary school places, and whose primary education had not prepared them for the available job or self-employment market. At the same time, Botswana inspired maverick South Africans, Brits and Batswana to start alternative secondary schools (the Swaneng Hill and Shashi River Schools),[6] to start the whole Brigade movement (see later), and to initiate the "Development Studies" curricula for schools.

The maverick South Africans were a valuable resource for Botswana: they despaired of the apartheid regime in their home country but were Africans first and foremost, looking to build, if possible, a way of life close to home, but very different from the hated apartheid system.[7] Such maverick South Africans, influenced, it is claimed, by Ghana, set up a series of new institutions called Brigades. These were secondary-level vocational training schools intended to be the seeds of local development initiatives throughout the country. The Brigades were, initially, at odds with the formal education structures of the colonial and now newly independent country. They were not backed by the government, and they developed as a variety of trusts, associations, NGOs and offshoots of local government.

They managed to survive and thrive, however, because the tribal structures of Botswana were very aware of large numbers of unemployed primary school leavers, and the need for such young people to get employable skills that fitted the local social and economic conditions. Many Brigades were encouraged by local government and local tribal councils that saw the point of experimenting and doing things differently.

In Maun, the local North-West District Council, supported by the Batawana Tribal Council, decided that they wanted a Brigade and negotiated with a British NGO to help them bring it about. They had, moreover, a resource they were prepared to commit: the fund that accumulated from impounded cattle, called *Mamelodi.* If cattle were unsupervised and grazed other people's crops, they were legally impounded by the district council and could be bought back by their owners on payment of a fee.[8] The funds thus accrued were the start of the Brigade of the North-West District, and this meant powerful ownership of the Brigade by the Batawana.

The British NGO brought me out to be the coordinator of the new Brigade, with a direct line of responsibility to the local district council. Other Brigades in Botswana were mostly in the industrialising east of the country, along "the line of rail" – i.e. the railway from Bulawayo to Johannesburg, through eastern Botswana. These Brigades saw themselves as producing young people who would get jobs in the new industrialising sectors, or who would make goods and services for the new middle classes. Maun's Brigades were meant to fit the area's own unique needs – far away from the development of the rest of the country.

And so, after checking out all the other diverse Brigades and their founders, we came to choose Building, Carpentry, and Arable Farming, and we went to the *Kgotla*, the tribal town hall meeting, to try to sell our ideas to the people of the tribe.[9] There was a lot of building going on, but no local building firms, and this seemed to be a field in which the tribesmen and women saw a future for their children. On top of that, all furniture came second-hand from the line of rail, and yet the North-West was one of the few parts of the country that had trees big enough to be sawn for timber – so people saw a future in this.[10] Arable farming was possible in the light sandy soils, and very different from the huge eight-oxen in-spanned ploughing of the rest of the country; most Batawana appreciated that farming on the receding floodplain gave them a comparative advantage for growing food crops, and were interested in donkey-drawn ploughing and weeding technology.[11]

We seemed to have the agreement of the tribe for the curricula of the Brigade, but then, after the *Kgotla* meeting, the Chief's wife came to see me and pointed out that we were not offering anything for young girls. She had just the teacher to suggest – a talented and disabled dressmaker who was local and could train young women to satisfy the growing need for school uniforms; they could also be self-employed to make local clothes as the taste for fashionable clothing expanded. And so the Brigade, now called NYTC (Ngamiland Youth Training Centre), had four trades to offer: arable farming, building, carpentry and dressmaking.

However, a young man had spoken up in the *Kgotla*, against the shushes of the elders, who gave very little space to young people to speak at all. He said, "I want to be an aircraft pilot. Can your new Brigade help me to be this?" He was told to be quiet and keep his place by the elders, but this man's cry stayed with me – we had not asked young people what they wanted. No, there was no way the new Brigade could teach him to be an aircraft pilot, but I thought that if we were offering ourselves as a way to train the future work-force of Ngamiland, we ought at least to have some role in career guidance for such ambitious youth.

The *Kgotla* gave us land on the other side of the river from the town of Maun, and we built simple dormitories, classrooms, workshops and teachers' quarters using mostly local materials.[12] The farm training was about 20 miles up-river on some land irrigated by flood water. We hired Zimbabwe artisans to be our teachers (but Zimbabweans who had been working in the North-West for some time) and a Peace Corps volunteer from the USA who was a farmer – there were few indigenous craft- or trade-trained people in Ngamiland – and helped them design lesson plans and curricula with a strong emphasis on learning by doing. The philosophy behind the Brigades was that they would be sustainable because they could pay for their running costs from the locally appropriate products made and sold during the training. It was a radical and exciting idea – and very different from the established way of trade training. Botswana had one national trade training school, linked to UK City and Guilds certificates, where young people were schooled in industrial skills that would lead to employment in factories. These schools, like their cousins in all ex-British colonies, were fully subsidised by the Ministry of Education and were islands in the places they were established – they had no spill-over influence on the populations or places where they were located. Brigades were meant to be places that would not only produce young people trained for existing jobs or self-employment, but would also act as engines of local development.

I was a strong advocate for this practical and pragmatic approach, but very few people had ever applied it before and there were few models to learn from. In the two- or three-year course, the intention was to produce all-round skilled artisans who could fit into the local market or go into business for themselves, and, while doing this, would produce goods and services that satisfied a local need and, most importantly, that people were prepared to pay for. We had to offer the NYTC as a commercially attractive and competitive workplace that would deliver professionally and also give its students well-rounded training. We did not want to have a lot of students condemned to mixing cement for two years while the teachers or the final-year students did the clever jobs like casting beams or fancy plastering.

The Building Brigade took on contracts for government buildings – clinics, teachers' quarters – but we were able to go up to only one storey.

We made buildings according to government standards and specifications, but on labour-only contracts – we did not have the capital to purchase building supplies. The trick was to get trainees from first-, second- and third-year entries to practise increasingly difficult parts of building. The Carpentry Brigade took on the construction of school furniture, roofing joists and, where we could get good local timber, tables and chairs. The Farmers Brigade grew maize, beans and peas, and also pioneered the growing of cotton. The Dressmaking Brigade made school uniforms. Of course, the more experienced the trainees were, the faster they could work. We also made sure that they learned costing and marketing. All trainees had to learn all the skills, and devising curricula and timetables for this was complicated.[13] We also taught Development Studies in the evening.

One fine tradition of Botswana was that the annual Agricultural Show circulated every year to a different district – and a year after the NYTC started, it was Maun's turn to host the show. All the local institutions and businesses made their own stalls to display their products competitively for a prize. NYTC decided to compete with a will, even though one of the competitors was the local taxidermist displaying a fully mounted and stuffed lion in an eye-catching set-up. The NYTC built a circular tower, showed off its chairs and tables, racks of school uniforms and groaning sacks of harvested maize and beans – and won the prize for best in show. One of the interesting offshoots of this was the comments of the academic secondary school students, who sometimes in the past had been a bit snooty about the trade training students: they were deeply impressed with what was shown – and said so to their peers.

Some snapshots illustrate the trials and lessons of the NYTC.

Classroom work

Even though the trainees were proud of what they were doing and producing, they still felt the gap between them and the students of the academic secondary school about a mile away, who had pens, exercise books and textbooks and sat at desks. The Brigades as a whole, and NYTC in particular, felt that they needed to temper learning by doing with learning through more traditional teaching, and so they set up evening classes. One part of the academic work was certainly connected to their trade – estimating, budgeting, costing – but we also introduced "Development Studies", making this a kind of sociological and economic look at where the country was going and what it was meant to be doing. We were helped greatly by the University of Botswana producing a popular version of the country's five-year plan; this was our basic text and was introduced as a kind of Socratic dialogue where the teacher asked the trainees for their thoughts about what the country was planning to do. The students were very keen on this and were quick to point out that the academic secondary school had nothing like it.[14]

Competition from the tourist trade

The most valuable sites for arable farming were those adjacent to the River Botletle, which was the exit for the waters of the Okavango Swamps as they came past Maun. Not so surprisingly, these sites were also very attractive to the growing tourist businesses, which provided accommodation in traditional *rondavels* (grass-roofed traditional houses) and trips into the swamps by dug-out canoe (*makoro*). After the tribal council had given us land just north of Maun for the farm training, we were told we would be taken to court for spoiling the amenities of a safari camp on the other side of the river.

A dapper (white) lawyer from Johannesburg arrived and required a meeting of the district council to adjudicate. The accusation was that our farming training site would become "a truck farm" (I presume the images came from South Africa, with broken-down tractors and decrepit buildings). The lawyer overstated his case and spoke as though Botswana was a Bantustan. No one was impressed and the case was dropped.

Raising money

The money from the *Mamelodi* fund had floated the start of the NYTC, and the projections that we made about selling buildings on contract, plus furniture, clothes and food grains, made the idea of financial sustainability look realistic, but we needed money to tide us over until the NYTC was in its second and third years and was in full production. Donors had to be found. We were an attractive proposition for most donors – we had huge local support and a long-term plan geared to self-reliance, and (not least) we provided an opportunity for donor staff to come to the beautiful Okavango as part of their visit to NYTC – but we fell into the interesting situation of being "too bush". I had deliberately built the physical plant of the NYTC (dormitories, canteen, washing and sanitation) to reflect local building standards, intending that the training centre would be seen as part of local life – and so mud bricks, chicken wire and plaster and thatch roofing were used. When I hawked my proposals along with drawings and photos around donors, I found that they were uneasy about this and wanted more concrete and more use of modern building materials before they would invest in the school.

Some of them were also hesitant about our objective of financial self-sufficiency, principally because there were no successful models for this anywhere else (the other model that I know of – Kenyan Village Polytechnics – were just starting around this time). They were prepared to allow the possibility of partial self-reliance, but were sceptical that this could be accomplished 100%.

Recruiting trainees

The *Kgotla* and the Chief sensibly suggested that a respected member of the Batawana tribal structure take on the responsibility of introducing the NYTC to more remote villages and small towns outside Maun. The Chief's uncle was given this task and travelled with us all over Ngamiland, talking about what was on offer and urging people to send their young people to enrol. The NYTC was free, as was the academic secondary school, but people in remote areas did not easily receive information about what was happening in Maun. The Chief's uncle was also very interested in a trip, with the transport paid by NYTC, since it gave him an opportunity for hunting game. Every Batawana tribesman had the right to a certain number of game animals under what was called a "pot licence", and he made sure he used up his quotient.

Government takeover

There were people in central government, particularly the Ministry of Education, who were unfriendly to the idea of a parallel educational model being promoted and implemented outside of government. A Brigades Coordinating Committee existed and liaised with government, but the Brigades had found their own funding and were run by people who were not civil servants, and not under ministry control. Maun NYTC's governance was clear since it was controlled by the North-West District Council, but the other Brigades had a variety of legal identities.

Thus there was a background at this time of the government wanting to take over the Brigades and establish control over this collection of disparate bodies. They did not fit government ways of working – straddling education, industry, local government and rural development. The Brigades had brought new funding into the country, which was valuable for a poor country, but Botswana was becoming richer every day and knew that it did not have to look for extra funds to manage these ideas.

What did I learn from the coalface?

The idea of linking production to vocational training, and having the former pay for a substantial portion of the latter, is a very powerful idea that can and should be tried in many different contexts. Many countries persist with expensive, government-subsidised trade training centres whose production is discarded as a by-product of the training process rather than used to either make money or link the training to realistic markets. After Botswana I think the idea has lost its adherents, which is a great pity.

The idea of developing a curriculum for secondary-level educational institutions called "Development Studies" and using it to prepare young people to think about the political and economic possibilities of their country is a very useful part of civic education and should be further encouraged. It requires an approach that is populist, not one that is exclusive.

It is very important to work with whatever democratic channels exist in any particular place to get local people onboard with an idea that may initially have only a few promoters. They are likely to suggest ideas that you had not thought of – like the introductions to remote villages through the Chief's uncle.

It is valuable to consider the situation of the people who are not going to be helped by the idea you are promoting. In this case, general career education for young people would have been a helpful complement to a particular educational track being proposed.

What happened to it all?

All the Brigades in Botswana were finally taken over by the government, but they were not turned into government bureaucracies – they still retained a measure of independence and creativity. As Botswana became richer, it was well able to pay for such institutions itself, but in so doing it lost the connection to the local economy and labour market.

There is now a government vocational training centre in Maun. By the same token, the vocational training centre was centralised to the Ministry of Education and no longer linked to the North-West District Council.

Many of the innovations in donkey-powered draught equipment for arable cultivation were taken up by local farmers.

Botswana is a role model for all of Africa for having taken riches from its diamond deposits and used them for investing in building the standard of living for its people. There are tarmac roads around the Okavango now, and Botswana provides free ARVs[15] for all its citizens who have AIDS.

Maun is now very involved in the safari tourist business, acting as the hub for a huge amount of popular tourism and game watching. At the time, game watching and safari hunting were very exclusive and very expensive pastimes for very rich people.

On a more personal note, my mother came from England to visit me in Maun at the training centre. She had my bed and I slept on a camp bed. She was a bit flummoxed by the latrine – "Where is the loo paper?" she asked. "Oh," I said, "pick a leaf!". She returned a bit later, saying that the leaves were rather small. She came with me and one of the American anthropologists and their backup crew of San. Their Setswana name for her was "Masadi Mogolo", which translated into English as "The Old Woman".

I also learned a great deal about the San – although not much that helped the Brigade. There was a huge amount of semi-mystical beliefs about the San among outsiders – that they were very spiritual, the first "men", the true spirit of Africa, etc. The only part of this that I witnessed was their astonishing night-time trance dances. After an extended period of rhythmic dancing around a fire in a sort of conga, supported by the singing of women, a man would go into a trance, cry out, and run off into the bush at speed. Others would run after him, bring him back and massage him. He would speak in tongues and slowly come down from his trance and drop back into the dancing.

When they camped at the Brigade, the San built their own fire and sat around it talking into the night. Given that in my American friends I had people who spoke the local San language (Ju/'hoansi – one of several San languages), I thought this was a great opportunity to listen in to the "spirit" of Africa. They talked of many things: what they had seen in Maun, game, and car mechanics – vehicles were literally "anatomised", with all their parts and functions given a parallel to the human body. There were also many jokes – one story, repeated many times, described an occasion nearby when a San had squatted to crap and a snake had slid out from between his legs. This was repeated frequently and lost nothing in its telling.

Another memorable feature of San life was that the Harvard anthropologists helped the first San acquire a driving licence, with the blessing and support of the Public Works Department of the Government of Botswana. He had been driving a Land Rover for a long time in the bush, but everyone felt that having a real driving licence would be a good investment in his future. He had to have a cushion to be able to see over the steering wheel.

There were a lot of elements of South Africa and apartheid in Botswana at this time. When Batswana addressed whites they used the term "*Morena*", which I thought was just the term for foreigner or white person, like "*Muzungu*" in Swahili. Then one day I was listening to a church service in Setswana and heard the word "*Morena*" being used a lot. I gradually came to understand that we were being called "My Lord" or "God" by the Batswana, which was sobering – and, for me, unacceptable. A continuation of this story is that the Peace Corps volunteer working for the Farming Brigade was a very large Illinois farm boy with a blonde Viking beard. He was also very unhappy about being called "*Morena*", and when this happened he made sure he voiced his objection, in Setswana. He said sternly, "*Ha ke Morena – ke mutu hela*" – "I am not *Morena,* I am just a man." The Batswana, who were already pretty taken aback by this giant, would say, "Yes, Yes, *Morena,*" and it would start again.

There were not many foreigners in Maun, and most of them were from the safari companies; they kept to themselves, and so I was considered available to the locals if there was any event that they wanted to make something special of. These included the regular beauty competitions in which the women and girls dressed up to compete for prizes, and I was asked to be a judge. At this time of Angela Davis, Afros and Black Power, black South African women would buy and wear imported Afro wigs over their African hair to look hugely cool. Batawana women in Maun wanted to do likewise. I made it known that I would only judge women with their own hair, which was beautiful without a wig. This was an unexpected answer, and I was not asked again.

When the town of Maun wanted to raise funds for public purposes, it had an amazing and, as far as I knew, homemade way of doing this.[16] A hall was booked, a choir was engaged, and the occasion advertised. On the due date there was a packed crowd and the choir started singing. One person, usually someone known locally as being quite well-off, had previously been prompted to stand up and announce that he would donate a sum of money to have the choir sit down, unless someone else outbid him. Once the choir was quiet, he or she would then identify someone else in the room and bid a certain amount of money for them to sing, sometimes specifying what they should sing. They could sing, or they could outbid the original person and nominate someone else – unless a third person outbid them in turn. Sometimes the singing was carried out (Batswana love good singing) and sometimes the strategy was to embarrass someone who was known to be a bad singer. If you had sung, you could nominate someone to sing next, unless they, or someone else, jumped in to pay money for them not to sing. And so it went on – pretty involved, embarrassing for many, but great fun, and it made lots of money for a local cause, usually to do with health.

What were we thinking at the time?

There were considerable debates going on at the time about Brigades and Development Studies, but most of it was informal and produced for contemporary workshops. The following were not available at the time I was there, but came soon afterwards:

Chilisa, Ernest. 1978. Botswana's Brigades provide practical training for development. IDRC Feature, International Development Research Centre (IDRC), Ottawa, Canada.

Graaff, J. F. 1980. The Brigades of Botswana. *Social Dynamics*, 6 (1): 25–35. doi:10.1080/02533958008458267

Patrick van Rensburg's comprehensive study came out in 1996, but unfortunately did not show how the ideas had been taken up by others:

van Rensburg, Patrick. 1974. *Report from Swaneng Hill: Education and employment in an African country*. Dag Hammarskjöld Foundation, Uppsala, Sweden.

van Rensburg, Patrick and Chiepe, Gaositwe K. T. 1996. Education with production: an overview [and] the Botswana Brigades 1965–1995. *Education with Production*, 11 (2): 75–90.

Many books have been written about the San, but Marshall Thomas's seminal work of 1959 was required reading for anyone interested in them:

Thomas, Elizabeth Marshall. 1959. *The Harmless People*. Penguin Random House, London, UK and New York, USA.

Then in 1982 came:

Katz, Richard. 1982. *Boiling Energy: Community healing among the Kalahari Kung*. Harvard University Press, Harvard MA, USA.

And finally, the astonishing story of one of the anthropologists whom I knew when I was there, Marjorie Shostak, who wrote *Nisa*, then, when diagnosed with breast cancer, came back to find "Nisa" (a pseudonym) and wrote *Return to Nisa*.

Shostak, Marjorie. 2000. *Nisa: Life and words of a !Kung woman*. Harvard University Press, Harvard MA, USA.

Shostak, Marjorie. 2002. *Nisa: Return to Nisa*. Harvard University Press, Harvard MA, USA.

Interestingly, another of the anthropologists whom I knew (Megan Biesele) moved from anthropology to social activism to get involved in the contemporary problems of the San.

Biesele, Megan. 2003. The Kalahari Peoples Fund: activist legacy of the Harvard Kalahari Research Group. *Anthropologica*, 45 (1): 79–88. doi:10.2307/25606115

Notes

1 Not the UN, but a small NGO.
2 In the apartheid era of South Africa (which was still in operation at this time), Bantustans were the South African government's attempt to demonstrate separate but equal development. They were a fraud.
3 The dry Kalahari around the Okavango Swamps is also where a small population of San people live – often referred to as Bushmen, which they dislike and see as a derogatory term. Through my friendship with Harvard University anthropologists who had been studying the San in Ngamiland for over two decades, I was able to become quite closely acquainted with them, dance with them, and see their very interesting interaction with the Batawana. The Batawana thought that the San could speak to birds and use powerful bush medicine, and the San were quite happy to encourage them in these beliefs.
4 The North-West District of Botswana was also called Ngamiland, and its people were a sub-tribe of the Batswana, called Batawana. They spoke Setswana, the national language, but had enclaves of other language groups: the Bayei, the Mbukushu and the San.
5 Other African countries that had discovered diamonds (Liberia, Sierra Leone, Angola) were no great advertisement – afflicted with what became called the "Resource Curse". Botswana was determined not to follow them and entered into a beneficial partnership with De Beers from South Africa. Botswana has become a model that other countries with mineral riches can learn from.
6 For many maverick South Africans, Botswana offered them a chance to do things properly in an African environment, and to escape from the despised and despaired of structure of apartheid. For other South Africans, Botswana was simply the fabled "bush" – the land of their *voortrekker* ancestors, where they could put on felt hats, drive Land Rovers and shoot game, and be allowed to do so by the government.
7 An added attraction was interracial marriage – outlawed by apartheid, but proudly proclaimed by Botswana, whose first president, Seretse Khama, married a white British woman. See the tremendous film about their romance called *A United Kingdom* (2016). They had to contend with both British colonial officers and their own tribal council.
8 These *Mamelodi* fees were collected at occasional sales of impounded cattle. The written notices for these identified each animal in Setswana and English and were a rare place to find words unfamiliar in English to describe cattle colouration, horns and dimensions, but which were precise and functional in Setswana.
9 The *Kgotla* is an impressive institution. Many wizened old men carrying their own traditional chairs arrive; women sit around the edges; and the Chief takes them through the subjects to be discussed. The level of debate is impressive.
10 I never saw a saw mill in the North-West at this time. Planks were produced by pit sawing, a very labour-intensive process.

11 In-spanning eight or more oxen to carry their possessions to the lands and then ploughing these lands was an important part of Botswana cultural life. Batswana from the south were amazed at the Batawana ploughing with donkeys.

12 When my friends among the Harvard anthropologists came to visit with an extended family of San people, the dynamics were fascinating as the local Batawana had never seen San close to before.

13 The builders got a contract to build teachers' houses at the government secondary school a short way downriver. It was very instructive to see the Brigade boys operating in the academic students' school. There was very little belittling of the artisan students by the academic students – in fact, there was quite a lot of jealousy over their prowess.

14 The copies of the popular version of Botswana's five-year plan had pictures of the capital Gaborone that showed streets and streetlamps. It puzzled the students when I explained that streetlamps were for finding your way around and for discouraging robberies. They were convinced that all people in the capital were rich and did not need to rob each other.

15 ARVs are antiretrovirals, the medicine to control (not cure) HIV/AIDS and keep it at bay.

16 A friend from Botswana, on reading this, said that this was actually quite common in rural South Africa.

3 South Sudan, 1973–75

Reconstructing the country in its one short period of peace

... In which the Department of Social Welfare of the new Regional Administration of South Sudan[1] *in the short period of peace between the First and Second Anyanya Wars builds facilities for young people in four towns – working with diaspora youth who come back to rebuild their country.*

South Sudan has been the site of so much civil strife, war, hunger, warring Generals and violence, over so many years, that many people forget that there was a period of peace between one war and another – roughly from 1972 to 1977 – and that many humanitarian and development organisations found the opportunity to be useful and to help, optimistically, to rebuild the country. I was approached by one such organisation to work there.

Almost as soon as the Sudan became independent from British rule in 1955, the South demanded liberation and secession from the North. The fighting that ensued coalesced around the Anyanya rebels in the South, fighting the army of the Sudanese government. It is often called the First Anyanya War and ended in 1972 when the Addis Ababa Agreement established the South Sudanese Autonomous Region, which would run itself through a separate legislative and executive body, with the soldiers of the Anyanya incorporated into the Sudanese Army and Police Force. The peace came about because of, on the one hand, a very costly war for both the Northern Sudanese government and the Southern population, and, on the other, forceful mediation by the All Africa Conference of Churches and the World Council of Churches. The peace, on paper, lasted a decade until 1982, but in reality had collapsed by the late Seventies as fighting between both sides increased – to be called the Second Anyanya War.

From 1972 onwards, aid agencies started pouring into South Sudan. Its governance was formed from a difficult mixture of the old British-influenced colonial administration, the Northern Islamic administrative institutions, Muslim commercial and trading institutions with economic control of many aspects of the country, the Anyanya army and its structures of the people who had stayed and fought, and the intellectuals who had left the country during the fighting but now came back again to set up a new government. It took some time for these disparate groups to coalesce around a government structure, and government posts were duly filled, but unfortunately underpinned with the same tribal rivalry between Nuer, Shilluk, Dinka and others that has bedevilled South Sudan ever since.

An aid organisation that now no longer exists, the International Union for Child Welfare (IUCW) in Geneva, was invited by the Sudanese government to assist with reconstruction work in South Sudan, particularly with reference to vagrant children. It recruited two Brits – first Beryl Knotts (a professionally trained social worker with experience in Brazil, Peru and, most recently, Biafra), who then recruited me. We were advisers to the Social Welfare Department to mount a programme to try to resettle and rehabilitate young people affected by the war. The programme would do this by reinvigorating a Department of Social Welfare within the Ministry of Health[2] of the new Regional Administration of South Sudan, and training the staff of this new Department for its programmes. The IUCW two-year foreign-funded project picked up the initial costs of this, on the understanding that the costs from Year 3 onwards were to be absorbed by the Department.

There was every likelihood that the people of South Sudan, faced with a flood of people from the humanitarian aid trade, many of whom came with a charitable mentality often derived from their church background, would become dependent on handouts and develop a relief mindset. A Sudanese Special Fund was set up to interview incoming aid agencies, and they liked the IUCW plans. We were strongly in favour of working to instil a more independent, self-help approach, while appreciating that the new Ministry of Health (and the even newer Department of Social Welfare) had few resources of its own in the short term and would need external help. Our approach was to recruit South Sudanese social workers returning from the diaspora and help the Department to train them. The first task was to design and then apply a survey of their home areas. Based on the results of this, IUCW and the Department of Social Welfare staff went to work to build multipurpose community development centres in the major towns of the South (Juba, Wau, Malakal, Aweil) and to design programmes there for rehabilitating the young people who were their priority. There were very many vagrant children in the towns of the South who had run there for shelter from the civil wars. They found ways of fitting into extended families, clan relations and tribal affiliations – often with great difficulty since everyone was in dire straits and could not easily accommodate extra relations.

Recruiting social workers from the returned diaspora was an amazing experience. Over the 18 years of the civil war, many South Sudanese families had been broken up, and many younger people and their families had escaped into neighbouring countries where they had received education according to the very many different systems and structures of those countries. All were very enthusiastic to come back to live and work in the South, feeling themselves truly South Sudanese, but forming one professional development organisation out of so many strands was complex. We had well-trained and committed people not only from a variety of tribes but also from a variety of academic and social backgrounds and languages – from North Sudan and Arabic, from Ethiopia and Amharic, from Kenya and English/Swahili, from Uganda and many different languages, from DRC Congo and French, and even some returning from the Central African Republic and French. On top of this they all had their own tribal identity and tribal languages. They same from Shilluk, Dinka, Nuer, Anouak, Bari and a variety of smaller tribes, and, further, on top of this was a layer of the inherited colonial religious affiliations – Roman Catholic, Anglican and Presbyterian – based on the specific missions that had been in South Sudan earlier.

This variety referred not just to the would-be social workers but also to our counterparts in the government. The main government officials with whom we worked were Anouak (from Ethiopia), Shilluk (who had stayed in the South), Bari (who had also stayed in the South – and a priest) and Nuer (returned from a PhD in Germany), while our Minister was Dinka (and had returned from an extended PhD in the UK). According to the inherited bureaucracy of the colonial service, staff had to have seniority based on their academic qualifications (our returnee from Germany had a PhD but it was in atomic physics); according to the dispensation of the new government, tribal affiliations had to be balanced among the staff; and according to Beryl and myself, we needed people with relevant practical experience and commitment. It was complex, as mentioned above, and a lot was not evident to us but decided within South Sudanese negotiations.

As soon as possible Beryl and I tried to ascertain not only the backgrounds and affiliations of the potential staff but whether they saw their future in the South in terms of peace and social development with young people. Most did, but one person we recruited was of another persuasion. As soon as possible he used his position to collect ivory in the south-west, take it over the border into Congo, swap it for guns and give them to the Anyanya. He never came back to become a social worker. For some people the war continued.

Many, however, were sound and committed to the work. They had personal experience of what so many young people had gone through, great empathy with their situation, a great delight in being back in their own country and being able to contribute to its rehabilitation. They also were happy to have a job and a salary, and were not infected with the corruption that soon became part of the senior levels of government exposed to too much aid money. We were also delighted to find so many competent and temperamentally suited women wanting to join. Nepotism was a much more difficult topic: every South Sudanese seemed to feel it was his duty to try to get advantages for his or her fellow tribesmen and women – not surprising after 17 years of war.

Our first task was to create the infrastructure within which we could work. We decided this would be multipurpose community development centres that could be used for vocational training, different kinds of classes, individual counselling and a variety of group meetings. Neither our budget, nor our inclination, nor local availability suggested hiring building contractors to put these up for us, and so we decided to go the self-help route. We bought modular metal frameworks from Nairobi and erected them on sites provided to us by the local municipalities (where such institutions existed) or simply in places that local communities agreed we could use. We put on the corrugated iron roofs, made the blocks, built the external and internal walls, fitted the windows, and plastered everything necessary. The meaning of "we" in this context was an essential element in our approach. The ideas were explained in the local communities whose members were asked to volunteer their labour; the social workers were instructed and trained in basic building techniques; and we recruited a British volunteer builder with experience in Latin America to help supervise construction.

Asking social workers educated in Nairobi and Kampala, for instance, to make concrete blocks, lay them and plaster them was very different from their expectations, but with our example, and with the mentality that we tried to inculcate – that this was what communities had to do for themselves because no one else would do it for them – social workers became successful builders *and* community organisers and got the collaboration of members of the urban population. Our sites were not in any way settled or traditional communities, but were often made up of displaced people who were desperately trying to find a home for themselves and their families in the new dispensation of peace and the cessation of fighting. As we discovered, it was difficult to do community organising when there is no real community.

Our social workers, apart from being asked to get involved in tasks that had never appeared in their curricula, were employed, were receiving salaries, and had definite and worthwhile work to do whose purpose they could understand. We were concerned that the flood of aid agencies and relief supplies did not subvert this. South Sudan became flooded with aid agencies' Land Rovers and Toyota Land Cruisers taking foreigners and fieldworkers all over the place expensively and unsustainably. We wanted to show that this was not the only way to work. We bought our field staff very small motor scooters (we may have overplayed the idea of smallness) with which they could visit the communities where they worked. Only one of these motor scooters was ever stolen (by the same man who collected ivory to exchange for guns), but they generally were not suitable for bush conditions. They were also a bit small for the large women social workers as they skidded round muddy corners, but the important principle to be inculcated was that the social workers were not privileged beyond the aspirations of the people among whom they worked.

The foreign advisers lived by the same standards of low aspirations. We built ourselves, in a very short time, a very comfortable local-type thatched *rondavel* with electricity, water connections and a flush toilet. This compared favourably with the UN staff, who waited for months before they could start work because they did not have the Western concrete block-built housing the UN demanded for them.

There were many diseases in South Sudan – bad for foreigners, but not good for locals either. Not infrequently our counterparts were stricken with malaria. Although they had more resistance than foreigners had, we often met our co-workers with yellow eyes, high temperatures and a marked lack of energy.

Once our multipurpose centres were set up, we moved on to the activities in them. There was counselling, family tracing, linking children to services that they needed, and liaising with the Ministry of Justice, the Police Force and the Probation Service over delinquent children. There were classes for mothers on nutrition and basic education, and from my side there were attempts to provide children with skills and tools with which they could earn a living. Apart from blacksmiths making frightening spear and arrow heads with especially vicious curls and curves so that they could not be withdrawn from flesh once they entered it and a few woodcarvers making neck rests to keep warriors' hair off the ground when they slept, there were few artisans in South Sudan, apart from farmers, cattle herders and warriors – and little "modern" production. By the same token there were few shops that stocked any kind of artisanal tools. For young children to be able to construct furniture for sale, they needed saws, hammers, braces and bits and adzes (in the absence of planes), plus tape measures, straight edges and T-squares. I was able to purchase the rudiments of these (e.g. saw blades) from local suppliers, and classes had to start by making their own tools.

It was always a problem to decide on the level of technology we should teach – should we give them modern toolkits and set up ways in which they could save or make money to buy them for themselves? Or should we consciously and intentionally help them to make inexpensive but out-of-date tools so that at least they could start production, leading to an income for themselves?

One aspect of our work in South Sudan where I was ignorant was mental trauma and illness. If I had been working there now, this would have persuaded me of a different approach, and had I appreciated more about this I would have modified my way of working. One of the administrative staff at the Ministry of Health attached to our work was very introverted, frequently incapable because drunk, and often did not come to work at all. When I asked about him, I was told by one of our programme staff that he had experienced terrible horrors during the civil war and that we should not expect anything more of him. This staff member, who was a priest, gave me a powerful talk about what people, including himself, had witnessed, and warned me to remember this background to all the people we worked with. Many of the men in the local communities were ex-Anyanya fighters, sometimes with very vivid bullet wounds and scars, but also mental trauma that could not be seen.

Not surprisingly, life was not very peaceful at this time, although a peace was officially declared between North and South. There was one short outbreak of violence in Juba while I was there, inspired by political disputes over the water of the Nile going through the proposed Jonglei Canal and to Egypt – and one of our official government counterparts was arrested and imprisoned. This resulted in a very heightened state of alert in all the foreign aid agencies, and their people started to make elaborate security precautions preparatory to possible evacuation. I was dismayed to find from experienced aid workers that protection and evacuation were available only for foreigners – if the shit hit the fan, foreigners would be rescued but not local co-workers. It never came to that, but this made me reflect on the aspects of loyalty and trust we aid workers expected of our co-workers, but which we were not prepared to reciprocate.

What did I learn from the coalface?

When working in a country where tribal loyalties are so important to people, it seems sensible to try to bring issues of identity and loyalty to the fore in discussions with local staff, rather than subsume them as being something foreigners cannot understand and should not try. It may be difficult, but it has to be attempted.

Extra effort needs to be made to communicate with the local population when there is no indigenous community – simply a mixed group of people living beside each other, often with vivid recent memories of violence. And help them communicate with their neighbours.

When a parallel economy is created by the aid trade, particularly in a post-conflict situation, it is important to decide the extent to which you will enter that economy. It is possible that your organisation will lose staff to the flashier organisations, but it is more important to have staff who are committed to the mission and who feel fulfilled in their work.

If you have a plan to help people, it is very important to talk through with them what they might do with the help they will receive from you, to make sure your plans for assistance are pragmatic and realistic in the context they know best. I did not do this enough.

As a backdrop to the development work that we undertook were the logistical and personal issues of living and working in South Sudan. In theory we could buy supplies from Khartoum and have them shipped upstream on the Nile on the antiquated ferry, but Nairobi was easier and more efficient. I and my Ugandan driver would drive our ten-ton truck to Nairobi on 18-hour non-stop trips, spelling each other because stopping off in Uganda was tricky and neither of us wanted to do that. Idi Amin's ill-disciplined army controlled the roads, and the country, and all we wanted was to get through as quickly and as safely as possible. I was hugely helped by my driver, who could speak many Ugandan dialects and jolly the army personnel along. We loaded up in Nairobi and drove back the same way.

South Sudan was a disease-ridden country. Not only malaria – which was pretty debilitating the first time you got it – but also bacillary and amoebic dysentery, bilharzia (schistosomiasis) and, if you had a local girlfriend, probably gonorrhoea. I also developed a rather unique complaint from bathing in the Nile – "otitis media", an infection of the inner ear.

The organisation I worked for stipulated that we had to go to Nairobi every six months for a full medical check-up and R&R, for which I was very glad, because they discovered bilharzia before it had really got stuck into my body. The medicine, however, was pretty gruesome. It was called metronidazole and was psychotropic, producing visions and nightmares. The UN doctor told me that he hospitalised people before giving them that drug in case they tried to commit suicide!

In Nairobi I had an operation on my ear where they cut open the eardrum, removed the gunk from inside and allowed it to heal by itself – warning me not to go swimming as I could drown through my ear!

What happened to it all?

Shortly after we left, the Department of Social Welfare engaged a Kenyan to continue the training of the social workers. The community development centres continued training youth and women in carpentry, metalwork, handicrafts, cookery, tailoring, childcare and nutrition with supplementary feeding. Nursery schools were introduced. Social workers skills were further developed in the new University of Juba[3] and on some short courses in Nairobi. By the late Seventies and early Eighties, life was disrupted again by the Second Anyanya War, leading up to the present. It has brought immense social dislocation and destruction to the South, now an independent country. This has been fuelled by immoral, cruel and destructive leaders driven by tribal loyalties, greed and corruption. Insecurity, criminality and violence are very prevalent, and very large numbers of the population have been driven out of their villages into IDP (internally displaced persons) camps – living artificial lives with no clear future. The town of Malakal, for instance, where we built one of our community centres, is flattened with no buildings standing, and it is likely that to a lesser extent this has been the fate of all of our work. The government is largely bankrupt and the excitement and optimism at the time of the Declaration of Independence has been forgotten. I suspect, but cannot confirm, that the government offers few services to its citizenry. Foreign aid workers are largely confined to their compounds because of security problems (mugging, stealing, personal attacks and rape).

It is very depressing talking to present-day aid workers, hearing their stories and thinking of the opportunities that were open to us in 1973. The lessons we learned and the experiences we had, however, are very relevant to all post-conflict situations, wherever they may be.

My colleague has retired, a few social workers we trained are still working in the South with the Department of Social Welfare, others have died – and others have migrated to new countries. My daughter recently worked with an aid agency in the new South Sudan, with young people in IDP camps. The dislocation of social structures occasioned by the war has messed up the lives of so many South Sudanese – they don't have villages, traditions or leaders (except military commanders). Ownership of cattle, so important to the social structures, is now by virtue of capture, and marriage, often linked to cattle ownership, is often replaced by rape.

But to go back to the times when I was there, there were some bizarre adventures, as you might expect, working in such an isolated place.

My sister came to visit while I was there and we teamed up with five others to drive the long (and more difficult) way to Nairobi through the Lake Turkana area direct into Kenya, avoiding Uganda. We were carrying some ivory bracelets carved in South Sudan in a box to mail them to a relative in the USA. At the Kenya border we were stopped by the frontier police and extensively searched. They found the ivory, which they insisted was illegal Kenyan poached ivory. We insisted that it was from the other side of the border. We were imprisoned in Lokichogio and a District Commissioner's court convened the next day. Our ivory was confiscated, but we were freed and allowed to continue.

On this journey we stopped at the last town in South Sudan, called Kapoeta, where a man greeted us with a matchbox. Inside he showed us some slips of panned alluvial gold, and asked us, "If you're going to Nairobi and coming back the same way, could you buy me a battery-powered record player?" I was not there for the transaction, but my friend took the gold to Indian shops in Nairobi, had it assayed, turned into money, bought the record player and delivered it back to Kapoeta.

While we were in South Sudan, a British animal-catching safari company pitched up from Uganda. This was a bunch of very intrepid guys who caught all manner of wild game and took them back to England for safari parks. My sister and I were invited to come along to catch a zebra. The whole operation consisted of a light aircraft in radio contact going up, spotting a herd of zebra, and, like a sheepdog, spooking them to run in a certain direction. Then we, in follow-up vehicles with radio contact, drove hell for leather over the bush into and among the herd, standing up in the vehicle with a long bamboo and a noose at the end of it. With luck we slipped a lassoed noose over a chosen animal. Then the vehicle was slowed down and the zebra manhandled into a large crate, in which he was brought back to their camp and freed inside a large enclosure. After some discussion the wranglers decided that the box was too small for the size of the zebra and he would be stressed. So we let him out and started the whole process again. Quite a way to make a living.

What were we thinking at the time?

Understanding the context for development work in South Sudan was largely formed by the opinions of anthropologists and missionaries who lived there before the end of the First Anyanya War – little else was written. The influx of aid agencies produced a great quantity of project documents with their separate attempts at documenting the local context. In my opinion, the best came from the Sudan Council of Churches, linked to the Kenyan Council of Churches. Largely, however, we observed, reflected on experiences in other countries (Biafra for Beryl and Botswana for me) and tried to design the most relevant programmes of work that fitted local conditions.

Huge amounts of literature have been produced on South Sudan since that time – much of it dealing with the humanitarian emergencies, and much of it in the form of proposal documents to foreign funders. Things became more polarised and more violent and peace seemed less and less possible.

Possible other reading could be:

DeMarco, Nick. 2017. 5 facts about development projects in South Sudan. Blog, The Borgen Project, 2 December.

Elmusharaf, Khalifa, Byrne, Elaine and O'Donovan, Diarmuid. 2017. Social and traditional practices and their implications for families in South Sudan. *Reproductive Health*, 14: 10. doi:10.1186/s12978-016-0273-2

Kon, Madut. 2015. Institutional development, governance, and ethnic politics in South Sudan. *Journal of Global Economics*, 3 (2): 147. doi:10.4172/2375-4389.1000147

Mbaku, John Mukuk. 2013. *South Sudan: Searching for a credible alternative development path.* Brookings Institution, Washington DC, USA.

Price, Roz and Ormert, Anna. 2017. *Youth in South Sudan: Livelihoods and conflict. Helpdesk report.* Institute of Development Studies, University of Sussex, Brighton, UK.

Notes

1 The new country of South Sudan did not come into existence until 2011.

2 It was actually the comprehensive Ministry of Health, Social Welfare, Culture, Sports and Wildlife Conservation.

3 Many thanks to Betty Katoro for this information. Betty, who was one of the IUCW staff, finally emigrated to the UK but has stayed in touch with the local situation and friends left behind.

4 LSE and Patchwork Community, 1970–76

Keeping in touch with the UK

... In which I receive specialised training for overseas work which also educates me about the UK before going abroad – also my intermittent participation in community living in the UK.

Having worked for three years in Ethiopia, and having decided that this kind of development practitioner work was what I wanted as a career, I appreciated that a close knowledge of Shakespeare, Beowulf and Chaucer was not going to cut it in that world.[1] I needed to get some solid practical training. The London School of Economics and Political Science (LSE) ran a one-year Postgraduate Diploma in Social Administration that had a UK-based option and an overseas-based option. I applied for the overseas option – we were a small, select bunch. The course was originally designed by the illustrious Professor Richard Titmuss, a prime mover in the creation of the post-war welfare state in the UK, to cater for people who had realised later in life that they had a vocation for social policy and administration, and needed a transitional course to train for it. The course had many interesting people who were moving into the social work sphere from a range of other backgrounds – civil engineering, business, the clergy and charities. It was intense – all holidays had to be spent on fieldwork. For those of us in the overseas option, there was no great chance of overseas fieldwork – so I was found opportunities in the UK, which showed me a lot about how things operated with respect to poverty and deprivation there. I was attached to a borstal in Suffolk, a community group in the Gorbals of Glasgow and a probation office in Brixton – and Brixton led me to work in adventure playgrounds and visit Holloway Prison and the Juvenile Court in Wimbledon, London. Finally, I spent time at the British Council's Audio-visual Training Centre in London.

After two terms of a three-term course in which I was enrolled, I was offered a job with the United Nations Association (UNA) in Botswana to manage a vocational training centre in remote Maun. For me this was just the opportunity to balance the very urban experience I had had in Ethiopia, but I certainly did not want to lose the course at LSE. I visited Professor Titmuss in some trepidation to ask for a leave of absence from the course for two years to go to Botswana, emphasising that I would come back to continue and finish the course. He was sceptical. "This has rarely worked in the past," he commented, "but I do not want to stand in the way of this chance. When you come back, come back in time to do two more terms to catch up." I did, and to my delight I achieved a distinction when I took the final exams. Having more field experience made my LSE studying more pertinent and valuable to me.

Part of my pride in what I was able to offer in the developing world was that I was not suggesting lessons that should be learned from the West's (or the UK's) experience, but cross-fertilisation between different countries in the South. I could offer suggestions from Botswana to Kenya, from Papua New Guinea to the Caribbean, from Sri Lanka to Pakistan. I never suggested that the experiences from the UK were directly relevant to the countries where I worked in the South. They did, however, have an effect on me – to compare these experiences with what I discovered in the South, and to calibrate my response to the extent and suffering of people in different circumstances. There was, to my way of thinking, little difference between unemployed families with little furniture living in empty cold rooms in the slums of the Gorbals and similar poor families in their huts who I subsequently saw in south-west Zambia, decimated by AIDS. There was also great similarity between efforts to get a community to work together and form a playground association in Brixton, London and as seen later in Grand Bay, Dominica.

Following the period at LSE, I kept in touch with my peers, including one in particular – Gregory Moore – who was doing the UK option. He was a Franciscan monk who was an imposing sight in his monk's habit among the jeans and anoraks of the other students. Gregory had a loose relationship with the monastery, which allowed him to pursue his enthusiasm for "squatting".[2] He was a strong believer in community living and started a short-life housing association called "Patchwork" shortly after he left LSE. I became an "overseas member" inasmuch as I came to stay in the different houses they had acquired when I was back in the UK between work overseas and participated in the community living. It kept me grounded.

The way Patchwork worked was to approach a big building company and offer to take over a house (or houses) that the company had bought prior to knocking them down and building something larger in their place.

It took time to get planning permission, and the building company had to provide security during this time. Patchwork offered to take over the building, use it as housing for people needing accommodation (homeless, those coming out of institutions – prison, hospital, mental hospitals), provided that would-be residents agreed to communal living, and then give it back to the building company some time later when planning permission was agreed. Typically, this was a year to 18 months. Patchwork also accommodated several social workers of different kinds to be resident there in a completely informal capacity. They were around, but they were there because they also believed in the value of community and its help to people who had been institutionalised or hurt. No one was aware of who was providing services and who was receiving them. The community life absorbed everyone.

I was thus exposed to social work problems in the UK in two ways: first, taking up the fieldwork placements of my LSE course (interrupted by two years in Botswana); and second, being an overseas visiting member of Patchwork.

My first LSE field placement was at a borstal in Suffolk, famous because of its illustrious alumnus, Brendan Behan.[3] It was run on tightly disciplined lines by a retired naval commander. I was invited to sit in with him on his daily "open clinic". The lads who were sentenced there were a combination of young criminals and people whose only crime was being caught smoking or dealing in marijuana. I was then urged to mix with the lads in open assembly, looking at TV, talking of this and that. "So, have you smoked grass, then, Sir?" I was soon asked. I was a student at LSE, for heaven's sake, and it was the Seventies. I did not think I should start my relationship by lying, so said "Yes", and tried to pass on to other subjects.

Next day I was up before the same naval commander who told me that he had heard of this conversation. "Had I smoked grass?" "Yes," I answered. "You realise that many of these lads are doing time for just that, and you are poncing about, boasting of it? You are out of here on the first train." And I agreed with him. The system was the problem, but there was no chance to discuss that.

My second placement for LSE was in the Gorbals Group, a group of Christian social workers who worked with the dispirited people living in that slum in Glasgow. Arriving in the dark early morning of winter in Glasgow after an overnight bus from London was pretty shattering. Drunks lying in the gutters and lots of evidence of what Billy Connolly[4] calls the Glasgow Pizza (i.e. bursts of vomit) all over the pavements as I walked across the Clyde from the bus station to the Gorbals. I visited the families living there with members of the

Gorbals Group – in bleak, stone-cold rooms with very little furniture and a terrible sense of despair. The four-storey Gorbals tenement blocks of monumental stone with little ginnels (alleys) and yards. The Gorbals Group volunteers tried to befriend families in need and help get them their dues from "the Welfare". The reality behind Billy Connolly's jokes was no laughing matter.[5]

My third placement for LSE involved a raft of experiences that stayed with me overseas – loosely based at the office of the Probation Officer in Brixton. Part of his bailiwick was an adventure playground – and I was introduced to the heavy ideology felt by locals there in respect of playgrounds. To the people in South Brixton who wanted an adventure playground, the use of plastic or metal preformed swings or seesaws or climbing frames was anathema, and quite probably a plot by the council to rid their children of initiative and creativity. Playgrounds had to be made of found objects and assembled by children with the help of a few, probably anarchist, playground attendants. I was astonished by the political enthusiasm that adventure playgrounds encouraged, and the community meetings called to administer and run such playgrounds were lively – to say the least.

The Probation Officer at the time was responsible for trying to prepare the ground for a client of his in Holloway Women's Prison[6] who was shortly finishing her sentence and coming back to outside life and, if he and she were not careful, the same environment that had encouraged her in the crimes of stealing and violence that had got her into Holloway in the first place. He said we needed to go and visit her in Holloway and plan for her release, and this we did. All men who think they understand, on behalf of their wives, mothers and girlfriends, sexist and misogynistic abuse should visit a women's prison. The shock of the wave of aggressive sexuality from the women prisoners as you, a man, walk past their cells is something difficult to prepare for. These are not passive people; they are rampant, violent women who have not had any sight of men for considerable periods of time, and who shout all the things that men on scaffolding, roadworks or building sites have been shouting at women for years. It is shocking. We finally got through the corridors to our client to find a cabal of women co-prisoners physically holding onto her to stop her from getting involved in any more violence which would provoke further sentencing and postpone her leaving date, and trying to stop those interested in teeing her up to do so. I tried jokes, but this was much too raw for banter. Once such topics as accommodation, possible jobs and signing on for the dole had been handled, the Probation Officer and I were quite relieved to get out without being seized, stripped and assaulted. Sure the women were behind bars, but the aggression was palpable.

And my last placement involved Wimbledon Juvenile Court, where I was attached to a young lad who was on remand for breaking open a telephone box to steal the coins. It was clear to me that he had done it as part of gang behaviour in his part of town, but he was assigned a young solicitor, fresh out of college, who felt that he could get him off the charge. Here I was exposed to the legal aid system, which allots a lawyer to you not of your choice and yet you can do nothing to change it. The lawyer explained to me that his proposed defence was to be based on the impossibility of having sexual intercourse with a chicken, for which he had found legal precedent. Just as a person could not physically shag a chicken, so a young man could not physically have been able to break open a telephone box, given its structure, and that was how he was going to defend him in the juvenile court. I never saw the result.

This period of nine months covering my first two terms at LSE before leaving for Botswana and the second two terms on my return were very educational (which was what was intended) and pretty sobering in terms of the fieldwork. Of course, Ethiopia and Botswana were very different, but what I saw in the UK was good preparation for anyone working for the poor and marginalised (our constant refrain) anywhere in the world. Learning the gap between what was actually happening and what the policy said should happen stayed with me.

Then, after Ethiopia, after Botswana, after South Sudan, there was the serial return to the Patchwork Community – to the new/old houses that the Patchwork crew knocked into shape to take all sorts of folks from all sorts of environments, and where they tried to make community living into a reality that was helpful to them. They and I were a crazy bunch of people, all highly motivated, all believers in the osmotic effect of community living on society's misfits. We were all just this side of the law, and many were proud of this. People moonlighted to take part in squats because it was important to show that there were too many empty houses and too many people without accommodation. But I was a sojourner, a temporary colleague, and as soon as I got a new job, I was off to live overseas. I did what I could while I was there, but it was clear that the others did so much more.

What did I learn from the coalface?

I was radicalised into becoming an enthusiast for squatting in houses that stayed empty too long in places where there were many homeless. I was educated about women's sexual violence. I also learned the interesting fact that borstals offer plastic surgery to men with antisocial tattoos (e.g. "Shit" and "Fuck" on their knuckles) so that they have a chance of getting a job.

I learned how appalling social conditions can make for depression and mental conditions and that this is difficult to escape from.

I learned that punk rock reflected the zeitgeist pretty accurately.

What happened to it all?

The government's Ministry of Housing saw the value of this kind of short-life housing and Patchwork was to some extent absorbed into the mainstream. Gregory Moore was sadly killed in a car accident, but his work continued.

Adventure playgrounds gradually lost their manic enthusiasts and manufactured swings etc. became the norm. Did this reduce the creativity of the children playing on them? I do not know.

The Gorbals slums have now largely disappeared. They were knocked down and replaced with something less monumental and more human. Many of the people from the Gorbals slums, however, were rehoused before this in purpose-built buildings on the outskirts of Glasgow (Easterhouse and East Kilbride) – many of which, in turn, became a lesser form of slum. In general, Glasgow, in spite of Billy Connolly's jokes, is a much more pleasant place to live in now and was selected as the European City of Culture in 1990 to many people's astonishment!

The Citizens Theatre, lying like a beached whale in the areas of demolition of the Gorbals, still continues strongly – a strange mixture of great theatre, Scottish Socialism and gay activism.

LSE's Social Administration Department is now the Department of Social Policy, offering a BA/BSc. Whether it encourages people who want to change careers in the direction of social work, I do not know.

Looking back on my time there, I appreciate it was a precursor to my later enthusiasm for social accountability (see Chapters 17 and 19). LSE taught you what the policies were and why they had been designed; the fieldwork taught you to see what was the reality on the ground; and the subsequent discussions helped you to think through the strategies and actions needed to improve things.

What were we thinking at the time?

The Postgraduate Diploma in Social Administration (Overseas) was an ongoing concern with its own curriculum, tutors and lectures. We overlapped with other faculties at LSE – anthropology, economics and development planning. It was, however, an orphaned course which has now finished, being absorbed into the MA/MSc in Development Policy and Planning.

In 1974, Kit Russell, the tutor responsible for organising the fieldwork, did a review of impact, but unfortunately it did not include the overseas option.

Russell, Kit. 1981. *Changing Course: A follow-up study of students taking the Certificate and Diploma in Social Administration at London School of Economics and Political Science (LSE) 1949–73*. London School of Economics and Political Science, London, UK.

In general, it seems to me, LSE staff at this time still thought that learning how the UK worked was good preparation for working overseas.

Exley, Sonia. 2019. Social administration revisited: traditions of observational field work and their value. *Social Policy and Administration*, 53 (1): 49–62. https://doi.org/10.1111/spol.12411

The idea of cross-fertilisation of ideas and practice South–South or South–North had not been incorporated into the thinking of the Social Administration course, or indeed LSE as a whole. This has now improved, as can be seen in the following article by David Lewis:

Lewis, David. 2017. Should we pay more attention to South–North learning? *Human Service Organizations: Management, Leadership & Governance*, 41 (4): 327–331. https://doi.org/10.1080/23303131.2017.1366222

Notes

1 My first degree at Oxford was in English Language and Literature.

2 "Squatting" was occupying unoccupied housing in the UK. The legality varied over time, but it was both a pragmatic response to empty housing and many homeless people, and also a strong political statement about the inequity that this should be the case. Many ingenious activists managed to get into such closed buildings and then establish right of possession.

3 Irish playwright and boozer, author of *The Quare Fellow*.

4 Billy Connolly is a Scottish comedian who grew up in the Gorbals. His jokes, widely available on DVDs, are savagely funny.

5 A strange coincidence was that my future wife, Clare Blenkinsop, subsequently worked as manager of the Citizens Theatre in the middle of the Gorbals, on Gorbals Cross, but was not there at the time I was. I ended up shifting scenery and pulling curtains in panto at the Citizens in 1977 while wooing her.

6 Holloway Prison was the largest women's prison in the country, situated in London and endowed as an educational institution by John Holloway (no relation of mine), who made his pile in patent medicines in Victorian times and bequeathed the strange mock medieval building to the nation.

5 Dominica, West Indies, 1976–78

Demanding assistance from the State or the joys of self-help

... In which the self-help scheme in Dominica coordinated and enthused people on the island to build and sustain their own infrastructure, eschewing government handouts.

By this time I had undertaken my specialist training at the London School of Economics and Political Science (LSE) and had come out with a Postgraduate Diploma in Social Administration (Overseas) – with merit. The Overseas Development Administration (ODA), precursor to the present DFID (Department for International Development), offered me a job based in their Barbados office as Social Development Adviser for the Eastern Caribbean. There were two of us, one man and one woman, and we divided the islands between us. We, and a bunch of other advisers in different fields, suggested to ODA and its local office in Barbados how British aid in the Eastern Caribbean should be spent.[1]

ODA's understanding of social development was limited. The position was generally thought of as one that would involve itself in youth, sports, women's affairs and possibly community development, but good governance and social accountability were nowhere on the horizon, much less anything that questioned the competence and integrity of the services offered by the governments. The other aspect of the job which rankled was that it was carried out by short "visiting fireman" visits to the islands by plane[2] and largely involved top-down suggestions linked to the available funds.

My patch was the independent islands of St Lucia, St Vincent, Dominica, St Kitts/Nevis, Antigua and, surprisingly, the Turks and Caicos Islands, still a British colony. My remit was to work with the governments of these small countries, while representing the British Government – the only time I have ever done so. While I was free to contact and work with local NGOs and CSOs, it was very much within the context of what was acceptable to the different Caribbean island governments. And these governments were volatile! Not only were they always apprehensive of being overturned in elections, but the scale of local party politics was small, domestic and almost parochial.

During my initial visits to the governments of these islands I discovered real interest in a comprehensive community development programme in the island of Dominica. Dominica, one of the Windward Islands, was anglophone – but only just; its closest neighbours were the French territories of Guadeloupe and Martinique, and the first language of most people was French Creole. Dominica, with a population of 75,000, was a very mountainous territory and the maintenance of its infrastructure was an ongoing problem. It had, at that time, well-distributed and isolated communities based on available resources like fishing and flat land, but its economy was largely based on plantation agriculture – bananas and citrus fruits. The communities were largely self-reliant (although all were dependent to some extent on the Dominican diaspora and remittances sent home from the UK, Canada and the USA) and had a strong sense of their own identity. They had village councils whose elections were keenly contested.

Much of the physical assets of the island that had been created by the government were in poor repair. There was a general feeling in the population that if the government had built something then it was the government's responsibility to maintain it. They had paid their taxes, and government should pay for the upkeep. The Government of Dominica was conscious of the cost of maintaining its infrastructure, and well disposed to the idea that the costs could be covered from some other source than its tax base, but it was apprehensive of trying to manage a large programme itself. Excessive rains, resultant landslides and very difficult terrain meant that local infrastructure was under great pressure. The Department of Local Government was well aware of Dominican citizens' local pride in their villages and their readiness to volunteer for local efforts to maintain and strengthen their facilities. This gave birth to the Self-Help Programme.

The Self-Help Programme prepared people for its implementation by pointing out in pamphlets and on radio that people *did not* maintain structures built by government but *did* look after things they considered they owned, or that they had had a hand in constructing. This was widely accepted and fitted with the indigenous self-reliant spirit of the people.

Equally importantly, the government was not apprehensive that a programme of community development through self-help would offer an opportunity to political opportunists to overthrow it, but saw that such a programme, encouraged and put forward by government, would increase its popularity.

The Department of Local Government, with its Permanent Secretary, Mr Joseph, set up the Self-Help Programme, we negotiated a three-year grant from ODA, and it was agreed with ODA that, since this was the biggest project under Social Development, it made sense for me to move full time from Barbados to Dominica to help in its implementation.

The first task was to announce and communicate the nature of the programme to its potential customers – the isolated and self-reliant village communities scattered around the island. The idea was well received in general and people were ready to volunteer their time and effort to improve the facilities of their communities, but the concept had to overcome some peculiarly Dominican concerns:

- Villagers were very clear that their contributions of voluntary labour were not to be seen as support for the government or the political party in charge of the State at that time.
- Villagers were very clear that, while they were prepared to offer labour and time, they were not prepared to offer financial contributions. They had, they said, paid taxes, and government had to provide what was needed that cost money – the cement, roofing sheets, windows, doors, etc.
- Villagers were also very clear that any self-help effort had to be for the community as a whole, or an important section of the community. Roads that went to a specific person's plantation were not to be built under such a scheme, whereas small bridges, culverts, community centres and repairs to clinics and schools were all acceptable.
- The Self-Help Programme was very clear that it would design the interventions, and make up the budgets, once these were agreed by the local communities. It was apprehensive of being seen as supporting local contractors and contributing to the polarisation of the wealth of local people. The Self-Help Programme therefore supplied an architect, a civil engineer and a quantity surveyor as needed. It was also concerned about potential corruption if it did not control the budgets.
- The Self-Help Programme did not set rigid rules for the self-help efforts, in the sense that they had to be managed through party political activists, or particular church officials, or any specific group. Any collection of enthusiasts in a village could put forward a suggestion for a self-help effort, the group would be checked out by the programme to see whether it was likely to be able to deliver the bodies to participate, and it would be responsible for

managing the effort through to completion. The staff of the Department of Local Government became largely subsumed into the Self-Help Programme and helped to liaise with any other department that needed to be involved – Water, Public Works, Agriculture, etc.

While the programme of voluntary efforts could be seen as an implicit (or explicit) condemnation of the government for having let slip the standards of maintenance for communal infrastructure, this rarely surfaced to the extent that voluntary labour was refused. The Self-Help Programme tapped into traditional and customary approaches to village maintenance and development and was implemented in an inclusive and festive atmosphere, even when the tasks were substantial – very much a Caribbean approach to communal labour, with joint food preparation, music, singing and drinking.

The only negative side – also very Caribbean – was the objection to what was seen by some as unwanted modernisation and the growth of "Babylon".[3] Dominica had its home-grown Rastas (called "Dreads") who were influential with young people and advocated a return to bucolic isolation – growing "ground provisions" (traditional vegetables) and weed and cutting off contact with the larger world, and particularly with the metropolitan countries. Their objections to the Self-Help Programme mainly served to stiffen ordinary citizens' enthusiasm for self-help and to reinforce their objections to the Dreads' ideological platform.

Concerning the administration of the project, the budget that ODA paid for was sufficient for the size of the programme and it was able to deliver the resources that were required to meet the citizens' aspirations, once they had been stimulated. ODA was flexible enough to respond to requests once the case had been made; they therefore provided a reconditioned ten-ton lorry as the programme increased in size and were prepared to help the programme with technical assistance.

Community Development programmes, from my experience, have the tendency to suffer the following problems – few of which affected the Self-Help Programme in Dominica:

- The programme becomes politicised and support for it becomes polarised into different camps – which may prevent other camps from enjoying its benefits.
- The programme may be seen and used as a source of income for people who have succeeded in "capturing" it.

- The programme may lose the enthusiasm of the community and their readiness to work for free because it does not deal with topics or tasks that are important enough to the community.
- The programme does not produce enough of real value for the communities involved, or the quality of the products is poor and rejected.
- The programme loses momentum as people lose the enthusiasm to maintain what has been created.
- The participants who are the intended volunteers refuse to work without pay.

In Dominica we required limited contributions from the communities – digging roads, ditches and culverts, collecting stones for foundations, "tarrish" (a cementitious aggregate that occurs naturally) and sand. We did not ask them to become bricklayers or plasterers (although some undoubtedly had the experience for this) because this would mean that they would lose an income opportunity. We hired skilled labour where we needed to. We always recognised their achievements through articles in the national newspaper or on radio. We always emphasised that anything created by the people was likely to be maintained by the people. This proved to be true.

The programme lasted three years and improved the stock of assets on the island, while making sure that they were maintained because the people had a stake in what they had built.

What did I learn from the coalface?

- Keep community development efforts away from national or local politics, which will try to take credit, particularly if the efforts are popular and well accepted.
- Try to make sure that community development efforts do not enrich different segments of the population, leaving out the citizenry as a whole.
- Make sure that community development requirements for voluntary labour do not exceed the capacity of the targeted population.
- Where possible, subsume traditional customs and behaviours into the programmes.
- Allow a variety of people or groups in a community to take the initiative for a community development effort – make it as inclusive as possible.

What happened to it all?

- Shortly after I left Dominica, there was an outbreak of violence from the army and the armed police, and the people's respect for and trust in the government decreased greatly. This also polarised people politically into different sides and decreased the opportunities for communities to work together on joint projects and programmes.
- Again, shortly after I left Dominica in 1979, it was visited by a massive cyclone (Hurricane David) which destroyed a lot of infrastructure and ravaged the banana and citrus plantations. This was repeated in 2018, and, with global warming, there is every chance that Dominica will be revisited by cyclones more frequently.
- What I hope I left behind in Dominica is that the citizens have a willingness and readiness to undertake communal self-help work for no pay. It must not be forgotten, however, that this was built on a substantial aid contribution from ODA (DFID). The extent that such contributions, for such purposes, will be a feature of DFID's aid strategy in the Eastern Caribbean is unknown to me.

On a personal level, the island of Dominica was unique in many ways. First, make sure that you're talking about the right place – Dominica, not the Spanish-speaking Dominican Republic, which shares an island with Haiti.

Dominica was very lush, forested and very well watered (it claimed it had 365 rivers). It had an astonishing boiling lake in the centre of the island, from which led many hot rivers and even one hot waterfall. Cold, dramatic waterfalls it had in plenty. It was great walking country, and the Dominicans were very proud of the island's beauty, spurning places like Antigua, to which it exported water. There were also black sand beaches that were a strange sensation, broken down from the volcanic rock in the centre of the island. It has "sold" itself as an ecological paradise and has created a niche but profitable market for ecological tourists.

Its old history was citrus fruit: there were many plantations of oranges, lemons and grapefruit in the hills, but these were now going back to bush and being replaced by bananas. It also distilled its own very strong rum (like many West Indian islands). It had its own version of Rastas from Jamaica – "Dreads" – who were very keen on living naturally, eating naturally and smoking naturally. They were not as peaceful as the Rastas, however, and murders took place, often influenced by the hot politics on the island.

I met my wife in Dominica. She was from the UK and had come out to meet a friend of hers for Christmas. She soon settled in and my Dominican friends approved and told me, "You have a heavy chick there, man."

At the end of our time in Dominica together we were getting ready to go to the airport when a hurricane warning was put out – Hurricane David, 1979. We reached the airport by car and the winds were so strong that coconuts were being thrown from trees all around us. We walked to take shelter with two Belgian nuns who lived about two miles from the airport and spent the next 12 hours under a table in their kitchen as the hurricane passed over us, breaking trees in its path, but stopping for two hours' calm as the eye went over. In about the sixth hour the roof of the house blew off and landed in a tree nearby.

Once the storm had passed and was heading to Puerto Rico, we helped the nuns clean up their house and then walked the two miles to the airport, waiting for some planes to come in on which we could hitch a lift. The first two planes were carrying reporters from North American TV who could not film very much since all the roads were blocked by fallen trees. The third plane was from Barclays Bank, checking on the safety of its money. The bank staff were also blocked from travelling to Roseau, the capital, by fallen trees, but gave us a lift to Antigua, from where we went to New York and met friends who had seen television coverage of the hurricane. No news had come out of Dominica so we phoned our parents and told them we were OK.

What were we thinking at the time?

There is a long tradition of community development work in British colonial history, often linked to food for work and to natural disasters. For example:

Hodge, Peter and Brokensha, David. 1969. *Community Development: An interpretation*. Chandler Publishing Company, Chicago, USA.

I was aware of this tradition from Botswana but found so many unique features of Dominican life that gave the Self-Help Programme its particular strengths, many of which were linked to the active local government structures. The Self-Help Programme produced many self-help manuals as well as technical leaflets for road construction, culverts, block-making, etc.

Dominica – and, I think, all the Caribbean Islands, except Barbados – is ill-prepared for natural disasters such as hurricanes. In Barbados there are old, stone-built churches to which people run, but this was not the case in Dominica. Added to this was the fact that the last hurricane had been at least two generations earlier – there were no oral traditions that prepared people for what to do, and the government had no disaster response or rehabilitation unit.

Notes

1 This was my one and only job for the British Government. They told me that if I wanted the job I had to sign up to the Official Secrets Act, which was enforceable for life! I wavered, but signed in the end. When I arrived in Barbados at the ODA office I asked to see some official secrets, now that I had signed. What I saw was cocktail party gossip – a big let-down.

2 An unexpected result of all this was that I got a stress ulcer from too much getting up in the very early morning to go to the airport in Barbados and fly out to small islands. My doctor said, "This is the Caribbean, man – you don't get ulcers here."

3 "Babylon", heard in the songs of Bob Marley, was the all-embracing word in the Caribbean for unwanted modernisation, capitalism and excessive wealth.

6 South Pacific, 1979–80

Appropriate technology, ideologues and small gains

... In which a US appropriate technology organisation sought partners to work with innovative ideas in Papua New Guinea (PNG), Solomon Islands, Vanuatu, Tonga and Samoa and then sought ways to bring the ideas into production.

After working in Dominica in the West Indies, I was offered a job as a consultant at an organisation called Appropriate Technology International (ATI) in Washington DC. ATI was the American equivalent of the British organisation Intermediate Technology Development Group (ITDG), formed to promote the ideas of Fritz Schumacher of "Small is Beautiful" fame. I found myself inside a world movement, that of appropriate technology (AT) – almost a world religion at this time as it had enthusiastic adherents in very many countries all over the globe. ATI was funded directly by Congress, which gives some idea of the importance that AT had in the USA at the time.

Wikipedia gives us today the definition of AT as being "a movement encompassing technological choice and application that is small-scale, decentralized, labor-intensive, energy-efficient, environmentally sound, and locally autonomous". Much of what I worked on was indeed of this kind (food processing and preservation, water pumping, building components and prosthetics or equipment for the disabled), but, as I discovered, AT encouraged a way of thinking which allowed you to dream about and conceptualise larger ideas of technological choice – particularly around energy (wind, solar and water power, and biofuels).

My job was to seek possible partners for ATI in Southern Africa and the South Pacific. I was called a "bird dog" by my American colleagues, as my task was to spot interesting projects that ATI could possibly fund. This introduced me to AT enthusiasts all over the world who had taken up Schumacher's ideas and run with them – often at great speed. I was allowed to join their chase in some of the most beautiful places in the world, in the South Pacific (PNG, Solomon Islands, Vanuatu, Tonga, Samoa).

Schumacher's thesis was that of an economist: in situations where labour was plentiful and capital was expensive, or where the cost of technology was beyond the means of the locals, it made sense to use technology that emphasised labour and avoided expensive capital that also required a level of technological sophistication not available locally. He called this "intermediate technology", pointing out that most technology that was being promoted globally was developed in the modern Western world to fit its balance of resources. It did not fit places that were remote, poor and dependent on their own resources, thus the need for an intermediate technology – some way between village technology and Western industrial technology. Those who added to his thinking emphasised the more flexible concept of appropriate technology – and the appropriateness was to be decided on the basis of local conditions and context.

AT attracted philosophical and political ideologues, sometimes displacing the technologists who saw what was technically feasible and made decisions on that basis. It also attracted Westerners (and those locals who had been exposed to Western thinking) who considered that they had a handle on what was locally appropriate. Its various titles reflected the different thinking – village, small-scale, decentralised, labour-intensive, energy-efficient, environmentally sound, locally autonomous, and also Buddhist technology. There was something for everyone.

What was very important to the AT movement (although sometimes forgotten) was that the technology should be context-specific – it should fit the situation in which the users lived. Thus, depending on the geography, the climate, the culture, the resource endowment, the level of education and the economy, a handpump with parts that could easily be replaced was more appropriate than a motor-driven pump relying on imported fuel; animal draught power for farming was more appropriate than a mechanised tractor; "barefoot doctors" rather than university-educated medical professionals.

AT does not necessarily mean simple technology – the amount of research required to make workable wind turbines or fuel-efficient stoves was and is considerable – but it does emphasise fitting the technology to the context in which people live. As Schumacher said, intermediate technology was "technology as if people mattered".

The enthusiasts came from many backgrounds and persuasions:

- There were those who were enthusiastic about historical technology that the Western world had moved beyond. Modern Western societies had moved into industrial technology, but "old-fashioned" technology

still made sense in a different context – particularly one that was isolated and poor and did not have access to imports that they could afford. Thus, ram pumps were more appropriate than motor-driven pumps in places where the geography was suitable, traditional medicine fitted better than a reliance on imported drugs, treadle power over electrical power, solar drying and preservation over refrigeration.
- There were those whose enthusiasm was for the use of renewable energy – falling water, wind, the sun, wave power. This was considered both for energy directly applied and as a means to create electricity which could be used for other purposes. Often this was connected to reducing the use of existing non-renewable resources (like charcoal), which led to a lot of work on fuel-efficient stoves.
- There were those who were enthusiastic about small business opportunities – how people could be helped to start enterprises without large amounts of capital through the use of microfinancing.
- There were those who were in revolt against an over-industrialised world – looking to the "back to the land" and "homesteading" movements of the American hippies.

A considerable number of the practitioners of AT were mavericks, and even eccentrics. I heard Fritz Schumacher say, to considerable acclaim: "People call me a crank. I like cranks – they are simple devices that cause revolutions."

My work took me to some extraordinary places. As a consultant I was not living for extended periods with people at the coalface – I was a visitor and hopefully a useful one – but I had the opportunity to understand the living conditions and the economic and social factors that influenced many different kinds of people and discuss with them the possible ways in which carefully chosen technologies could help them.

Here are some of the contexts with which I was faced.

Papua New Guinea and ethanol from biofuels

PNG provided extreme examples of people living in isolation, far away from modern technologies, particularly if such technologies involved fossil fuels to power machinery. The cost of transporting oil to remote communities meant that oil-fired electricity plants were out of consideration for most people in most places. Although there was huge demand for electricity for light, heat and driving machinery, the cost of transporting fuel made it almost impossible, except in the enclaves of very high-technology mining towns. However, at the same time, Brazil was producing ethanol from sugar cane and powering cars and other machinery with it and encouraging others to consider the concept of biofuels. PNG did not have the kind of geography and climate to allow it to emulate Brazil, but the idea of producing fuel from the crops they had was born in the minds of some people in PNG.

With the help of US scientists they looked at examples of fuel from crops and saw the unlikely but possible example of learning from those who were presently producing ethanol from maize in the Midwest of the USA to drive their tractors. These people in Kansas were the grandchildren of the bootleg alcohol producers of the prohibition days. They were cooking maize, distilling ethanol and, with some adjustments in the carburettor, running a variety of internal combustion engines. I went to Kansas and interviewed some of these farmers and asked them what would be needed to set up a biofuel operation.

PNG did not – and still does not – have large fields of maize, like Kansas does, but it did have a substantial acreage of cassava, since cassava was the main food of the country. Biological scientists suggested that cassava could be cooked, a bit differently from maize and sugar cane, and could also produce ethanol, and ethanol could be used as a combustible fuel to drive electricity generators in remote, isolated areas. In theory, small-scale ethanol plants could be set up all over PNG to provide fuel from the surrounding agricultural areas. In fact, large-scale plants have been set up, such as the PNG Cassava Biofuel Project.

Solomon Islands: growing vegetables in the rainforest and moving on from shifting cultivation

In the Solomon Islands, the indigenous vegetation is tropical forest with huge trees that shade the soil, but soil with few nutrients that cannot easily raise any other kind of vegetation. There are also torrential rainstorms that erode any land not protected by tree cover, and fierce sunshine which scorches any plants unprotected by tree cover.

The people who live there practise cyclical cultivation of the forest (shifting cultivation – called "slash and burn" in some places) by selectively cutting down the trees in a chosen area (and leaving many standing), burning them (and using the ash as a fertiliser), digging over the exposed soil with digging sticks without disturbing the soil too much, and planting their food crops between the trees, taking crops from one or maybe two cycles of cultivation, and then moving on to an adjacent stand of tropical forest and doing the same. It is very arduous work. The crops they plant live because the soil has been fertilised by the ash of the burned trees and because the power of the burning sun, which would otherwise oxidise the soil and burn up the new shoots, is held back by the shade of the selected trees that are left standing.

In other parts of the world where tropical forest is cleared it is either replaced with other tree crops (like palm trees in Malaysia) or slowly becomes oxidised and barren (as in Brazil). The small village of Irriri, on Kolombangera in the Northern Solomons, with advice from an Australian AT organisation called APACE (now no longer in existence), decided to try something else. It clear felled a few acres to make a garden, thus exposing the soil to the sun and rain, then purchased a mulching machine powered by a small petrol engine. The vines and undergrowth that were available from clear cutting the forest were fed into the mulcher and large stockpiles of material produced to make very large compost heaps. Small tree trunks were used to outline and border garden beds to hold the soil in place under tropical rainstorms, and the food crops were planted in raised beds covered in large amounts of composted mulch which shaded the new growth from the sun, protected it from the rainstorms and fertilised it. The shoots were protected and grew up through the mulch.

Irriri was a Seventh-day Adventist village and was very active in trying out new ideas. Fuel for the small petrol engine to drive the mulching machine was available since supply lines for outboard engines for their motorised canoes were already in place. The village was thus enthusiastically changing from shifting cultivation to sustainable vegetable beds through the availability of very large amounts of composted mulch which had been produced by a relatively cheap and simple two-stroke petrol engine powering a mulcher, and the villagers, well practised in outboard engines, had little difficulty in maintaining the engines.

PNG again: airlifting oxen trained to plough

Traditional cultivation and planting techniques in PNG were based around the digging stick and heaped seedbeds to stop soil erosion. In some of the flatter lowlands, however, ox ploughing had become common and locally made ploughs as well as imported simple iron ploughs from India were being used. Much of PNG, however, consists of settlements that are very far apart from each other, and although most of these settlements are on mountains, it is not unusual to find areas of flat land or plateaus on the mountain tops which are very suitable for ox ploughing.

A local agricultural training centre called Yangpela Didiman (Neo-Melanesian for Young Farmer) provided agricultural training to many young men from the hills at their centre in Banz. A big part of their training was teaching ox-drawn ploughing and cultivation. Many of the young men from the mountains were envious and wanted to get oxen back up to their homes, but the jagged territory and distances meant that oxen could not walk to such settlements.

In PNG, many high and isolated settlements have small bush air-strips, put in originally at the request of the missionaries who wanted to bring in bibles and preachers to convert the PNG people. There was a well-organised system of airplane communication called MAF (Mission Aviation Fellowship), which had small planes, intrepid pilots and the infrastructure to link the planes on the lower areas to the air-strips in the high mountain villages. Putting it all together, Yangpela Didiman helped mountain farmers choose and train an ox for their own village and then transport it in a small plane to the high and isolated villages, where it could live and help develop sustainable and increased food production.

Biogas all over the place

Many institutions experimented with producing methane from the anaerobic digestion of vegetable material, often dung, slurry or human waste. I saw attempts at this in Tonga, Samoa, Vanuatu, the Solomon Islands and PNG. The reputation of Gobar Gas plants from India was very high, but the construction (they needed well-sealed components to make them anaerobic), management (they needed the correct proportions of organic material and water to produce useable amounts of gas) and burners (which had to be "tuned" to the correct air/methane mixture) often meant that their performance belied their reputation. It sounded so satisfactory – producing gas for cooking from waste and saving on firewood or charcoal, with the additional benefit of reducing health hazards of burns and lung problems for the women cooks – and yet at that time it was not so easy to establish a working and sustainable system.

Micro-hydro – also all over the place

Channelling falling water into a turbine which could then produce electricity was another idea that had many adherents but did not live up to people's dreams, particularly in PNG where there were many isolated communities with strong streams and waterfalls, but not too many ideas about what to do with the electricity generated. In some cases, watermills were used for direct power to grind corn or other agricultural produce, but once the idea was extended to produce electricity, then the turbine, even if taken from a recycled car, and the requirements for bearings and power distribution from the turbine to the settlements meant that the micro-hydro experiments provided opportunities for the burgeoning number of AT centres to practise their ideas but did not result in many improvements in the lives of the people. I remember an amazing water turbine harnessing power from a powerful canalised river in a remote part of PNG, but in the end all it did was produce hot water. This was enjoyed by the locals, but fell short of the ambitions of the AT ideologues.

What did I learn from the coalface?

For me, the most important aspect of encouraging and supporting AT, at that time, was that it encouraged and allowed ordinary citizens to have a transformational change in their thinking. Technology was not just something that came from towns and from imports – ordinary people could learn how to make substantial changes in their lives, using ideas that they would rarely have dreamed of. The ideology of AT was a powerful one that was empowering to many in the South – even though it was often taken up by overenthusiastic mavericks and eccentrics from outside the communities that would use the results of the technology. Schumacher's words were prophetic – cranks produce revolutions – but there were, in my experience, too many cranks, and the revolution was still far in the future.

What happened to it all?

Ah, the revolution did indeed happen, but it was far from the original definition of AT – "a movement encompassing technological choice and application that is small-scale, decentralized, labor-intensive, energy-efficient, environmentally sound, and locally autonomous". We now have many examples of very large, industrial-scale investments that arose from AT trials – huge methane gas plants consuming city waste, many examples of ethanol production from cassava in PNG, and, of course, a range of hydroelectricity plants, from micro-hydro to mega-hydro. As far as I know, trained oxen have established themselves in isolated communities in PNG, and Yangpela Didiman still trains young farmers to use them.

The technology that I have not seen develop further is the sustainable cultivation of vegetable beds in exposed tropical forests through using the protection of mulch produced from the waste vegetation of forest clearance. It is common to have plantation cultivation of coffee, for instance, protected by planted shade trees, but standard textbooks suggest improved "slash and burn" cultivation rather than intensive compost making.

And on a personal note

My year with ATI looking for partner organisations in the AT world in Southern Africa and the South Pacific gave me one of the longest airline itineraries I have ever had: USA, UK, Mauritius, Australia, PNG, Solomon Islands, Vanuatu, Tonga, Samoa, Hawaii, USA. Being somewhat ashamed of costing so much money and using so much fuel, and being an AT ideologue myself, I decided to buy a Moulton folding bicycle to travel with me, and at each airport I would unfold it and bicycle to the nearest town. Great in theory, but the distances, the time and the chances of getting lost defeated this idea – I only managed it in Samoa.

I spent three days in Mauritius, principally because of some airline difficulties, but took the opportunity to look for examples of AT. I was delighted to discover that much of the island's sugar cane was processed in large industrial plants that were fuelled by the waste product of the process of extracting the juice – the bagasse.

In the New Hebrides I was lucky enough to be linked up to Ken Calvert on Tanna Island. He is a New Zealand pastor and alternative energy expert who subsequently, with independence, became the Energy Minister of the new government of Vanuatu.

Tanna had the extraordinary "Jon Frum" cult. It is supposed (it's difficult to get clarity on this) that Tanna Islanders, amazed by the arrival of gigantic US Navy vessels and their cargo in the Second World War, and equally amazed that there were black American sailors, persuaded some to jump ship and stay with them. The Tanna Islanders believed that such people knew the code by which they could phone the White House and the President and get more ships to come to Tanna, and simulated telephones were strung up in coconut trees to do this. It is a more specific form of the "cargo" cults that were seen in many parts of the South Pacific.

One of the strangest sights in Tonga and Samoa were the Mormon missionaries. Somehow the Mormons (the Church of Jesus Christ of Latter-day Saints, headquartered in Utah, USA) had come to believe that the people of the South Pacific were a "lost" group of Mormons who needed to be reclaimed (not so different from the Falasha Jews of Ethiopia in Chapter 1). So you would see, in pairs, identically clothed young white men in short-sleeved white shirts and ties, walking around the islands preaching the message of Joseph Smith.

Fiji produced many varied experiences – for me as a Brit, one of the strangest was meeting Fijians who had enrolled in the British Army and had fought in Cyprus, Northern Ireland and other places. After they had finished their stint, some stayed on in the UK, but those who returned were equipped with very specialised knowledge and experience (and a range of British regional accents depending on the regiments they had been attached to). The ones I met were thinking of opening small businesses in the towns of Fiji and were the focus of the government's entrepreneurship development programmes. I subsequently met returned Gurkhas in Nepal in roughly the same situation. Later, I also met Fijians as peacekeepers with the UN in East Timor.

What were we thinking at the time?

The field of AT was heavily written up at this time, and I was an avid reader, particularly of the *Whole Earth Catalog*, the bible of the US hippie "back to the land" movement, publications from VITA (Volunteers in International Technical Assistance), and the booklets of ITDG. Even the South Pacific had its own AT bible – *The Liklik Buk* – produced in PNG. My major influences were, and are:

Brand, Stewart (ed.) (with others). Various editions (1968–98). *The Whole Earth Catalog*. Portola Institute, Menlo Park CA, USA.

Brand, Stewart (ed.). 1986. *The Essential Whole Earth Catalog: Access to tools and ideas*. Doubleday, Garden City NY, USA.

Koian, Rosa (ed.). 2003. *The Liklik Buk: A sourcebook for development workers in Papua New Guinea*. Liklik Buk Information Centre, Port Moresby, Papua New Guinea.

Schumacher, E. F. 1979. *Small Is Beautiful: A study of economics as if people mattered*. Harper Row, New York, USA.

Schumacher, E. F. and Gillingham, Peter. 1980. *Good Work*. HarperCollins, New York, USA.

But, as I have mentioned, AT was a complex topic. In some places and by some people it was considered "a poor man's solution", which did not help its adoption. It required policy environments which would support it and these were not forthcoming: development fashions morphed into support to free markets, which did not include reference to AT but access to cheap capital. The original organisations promoting AT changed (ITDG became Practical Action and ATI became Enterprise Works Worldwide/VITA), but the ideas evolved with the growth of the Open Design movement and Open Source technologies.

7 Java, Indonesia, 1979–84

More AT ideologues and people's technology

... In which Oxfam works in Central Java, Indonesia with villagers to deliver clean water, rear goats, and turn human waste into fish to eat and sell.

Following Hurricane David in Dominica I was recruited by Oxfam, which placed me as their representative in Java from 1979 to 1984 – not, as so often happened with international NGOs, based in the capital city, Jakarta, but based in Semarang, the capital of the province of Central Java, and covering Central Java and the special district of Jogjakarta, which is situated in the south of the province. There had been only one Oxfam rep before me, and so the interpretation of what Oxfam might do to implement its objectives of the time ("helping the poor to have more, and to be more") was wide open. In general, Oxfam encouraged its staff to identify motivated groups of people from poor and deprived backgrounds and help them think through what they could do to improve their lives. In Indonesia at the time, the government fully expected Oxfam and other international NGOs (INGOs) to support government programmes, and they were puzzled when we suggested other ways of working, and perhaps suspicious that a *non*-government organisation was actually a cover for an *anti*-government organisation.

At the time, the Indonesian Government was very enthusiastic about its Master Farmer programmes in which selected better-off farmers were helped to access and cultivate Green Revolution crops (mostly rice), together with the linked programme of pesticides and fertilisers. It was planned that such progressive farmers would demonstrate greater yields and persuade other farmers to follow their model. The rationale for this programme was to increase the national yield of rice because the population of this very large country was a net importer of rice, and it was a matter of national pride that Indonesia should be able to feed its own people.

Java is one of the most densely populated places in the world and every inch seemed to be under cultivation, with many people working as labourers on richer farmers' fields. Greater yields for landowners did not translate, however, into greater wages for labourers, and Oxfam found one of its niches in Central Java to be working with landless labourers to help them improve their wages, develop very small plots for themselves, join others with small plots into cooperatives, and try to get good prices for their produce.

When discussing development strategies for agriculture with government officials, and trying to persuade them of the value of working with the poorest (i.e. the landless labourers), the response was usually that such people would not contribute to the increase in the national yield, and sometimes that such people were not worth helping because they were too poor and could not respond to the incentives of the government. Indonesian economists at the time were producing evidence of the very large number of people who were landless labourers and thus outside the benefits of government programmes; this helped, but there was a suspicion that foreign NGOs were possibly aiming to return Indonesia to the days of the Communist Party – the civil strife of 1965 against the Communists that had resulted in so many deaths had occurred only just over 15 years previously.

Very frequently in Indonesia, local CSOs had very good ideas for development among the poor but found that the promotion of these ideas would be seen by the government as a challenge to their own programmes, and thus were to be treated with caution, if not prevented altogether. The government had a Department of Cooperatives, for instance, based on good principles, but with a very poor record of delivery. CSOs, which were well aware of the value of cooperative principles and practice among small farmers, found that they had to consciously design their work to avoid being seen as challenging government structures. Thus, one of the largest CSOs, Bina Swadaya, which worked to organise small farmers and the landless, called its groups "*Pra-Kooperasi*" (i.e. "pre-cooperatives") and if challenged by government officials would say that they were preparing the people to be able to join government cooperatives.

A way round these unhelpful government practices was to work not with the ministries of different sectors but with the decentralised Ministry of Interior and the local officials who had authority over a certain geographical area. These were, in order of importance: Governors of a province (*Gubernor*), dealing with populations larger than many countries; Regents (*Bupatis*), still dealing with several million people; and Sub-regents (*Camats*). These government officials were quite competitive and always seeking fame and reputation for their territories – and they were even more eager to make sure that all was seen to be peaceful and that there were no squabbles, outbreaks or disputes which would disrupt the harmony of the State, something that is very important to the Javanese in particular. Indonesian CSOs who had good developmental ideas tried to "sell" these to the government officials so that particular regions could gain a good reputation for development achievements.

If an idea from a CSO was accepted and promoted by the regional official, the sectoral ministry officials would get behind it and try to get some of the reflected glory. Many was the time that a project or programme organised by a CSO found government officials participating in the public launches or yearly anniversaries, claiming that their encouragement had made it all work. A very important word for the government was *pembinaan* – "guidance". The provision of the government's "guidance" gave them a stake in the CSO's work, without actually having to do anything else. This was the context in which Indonesian CSOs worked, and in which INGOs were able to support them and justify the resources used.

The organisation with which Oxfam Indonesia worked the most while I was the Country Representative was Yayasan Dian Desa (Lamp of the Village Foundation). Its Director was very smart in the way he related to local government: not only did he play tennis with the Governor of Central Java and liaise diplomatically with the *Bupatis* of the areas where we worked, but he was also given the Ramon Magsaysay Award from the Philippines (also called the Asian Nobel Prize), and this was noteworthy enough for him to get a meeting with President Suharto. This allowed him a photo opportunity (and the resultant photographs) that impressed all in government.

Yayasan Dian Desa and its Director, Anton Soejarwo, were deep believers in appropriate technology and equally deeply involved with village development. Their initial main claim to fame – and the way in which Oxfam became involved with them – was their development and manufacture of ram pumps. Ram pumps are water-delivery systems that pump small amounts of water uphill powered only by the pressure of large amounts of water falling downhill, channelled onto the pump and creating air pressure. They are mechanical but not complicated – well within the manufacturing capacity of small workshops and the maintenance capacity of villagers. In Java, the poorer people lived up the mountains where there was good cultivable soil that had not been taken by large farmers. Their water came from mountain springs, but these were not necessarily close to the villages, and so the water had to be carried long distances, often in long lines of hollowed-out tubes of bamboo snaking across the countryside, or in ceramic pots on the head. In some places there were astonishing long reticulated systems of bamboo piping crossing the countryside to bring water from the springs to households; the bamboo pipes had the nodes knocked out and the thinner ends were rammed into the next pipe's thick end, with holes drilled in the bamboo occasionally to equalise air pressure.

Yayasan Dian Desa, however, introduced ram pumps set up at the base of a chute of falling water channelled from the spring, and these led into a series of plastic (or bamboo) pipes (depending on budget) that fed the water uphill and across country to the villages. The ram pumps worked through the pressure of the falling water ramming a smaller amount of water under pressure into a small-diameter pipe which then carried the water large distances and up gradients. I think its provenance was an old technique for watering sheep in high pastures in Wales. The pump was assembled from simple iron components in Dian Desa's workshop together with gaskets formed from old rubber tyres. Villagers who wanted a ram pump had to accept responsibility for manhandling the pipes across country, digging and building the collecting basin and feed channel from the spring down into the ram pump, and then, at the other end, building a reservoir tank to hold the water from which the villagers could easily collect it. They also had to learn to replace the rubber washers from time to time, and commit to doing so. Where a ram pump had been installed you would hear a clicking sound at one-second intervals as the pump shot charges of water up the hills and into the reservoirs. You would also see small serial fountains of water squeezed from the rubber gaskets at each pulse of the pump.

Oxfam supported Dian Desa to place four or five pumps in villages in Central Java. All went very well and all the villagers were very satisfied. Dian Desa, however, was constantly striving for new ideas and new opportunities. They identified one settlement for a project big enough to need a great deal of water and a powerful waterfall that could feed water to a very large ram pump. They asked whether Oxfam would be prepared to fund it, and mentioned that the amount of water would be so great and the pressure on the pump so high that they would need specially reinforced piping. I was concerned that the delights of manufacturing this amazing technology, rather than the benefits to villagers, were starting to drive Dian Desa and so we agreed to postpone it. Some time later Dian Desa contacted me and said that they had managed to get pipes used for drilling for oil (cheap) and these pipes were strong enough for what they wanted. I agreed and they went ahead.

When I finally saw the installed water pump it was a giant – perhaps the largest ram pump ever manufactured. Where a usual ram pump pulsed every second or so, this massive machine worked to an eight-second cycle, building up the pressure of the falling water until it reached the required level, spray cascaded from the rubber joints and the water shot up the delivery pipe with a clang that echoed up its track. An amazing machine that worked very well and delighted the AT guys at Dian Desa enormously.

There were two other kinds of memorable projects that Oxfam invested in in Central Java – goat "hotels" and loos over fish ponds.

In the case of goat hotels, we were concerned about helping villagers build up some assets against the usual problems that affected poor people – sickness, accidents, paying for school fees, paying for other government charges, drought, floods or any other kind of distress. Goats, for those who had them, were normally corralled at night beside the house and taken to be staked out for feeding near the house every day. They were often in danger of being attacked by dogs or stolen by people, and they needed to be watched over. We saw that some smart villagers kept their goats in stalls and cut and carried grass to them every day. Goats were hardy, bred quite easily, and had a commercial value as they could be sold and converted into needed money without much difficulty, and without dropping villagers back into poverty. We helped villagers to build bamboo sheds for their goats off the ground, cut and carry grass to them, and collect their dung easily from below the "goat hotel" to fertilise kitchen gardens. The goats had clean stalls (dung fell through the cracks), they could easily be watched, and they bred and multiplied. An extra element was veterinary training for goat hoteliers so that they could keep their animals healthy.

Fish pond latrines were a way of earning income from raising fish in fish ponds and selling them. There were many places in Java with clay soils that allowed villagers to build fish ponds with access to dirty surface water run-off from the rain. Villagers were also accustomed to raising fish and many of them had their household latrines built on the banks of the pond, letting waste flow into it. A few of them had created communal village latrines. This is what we promoted. Each village would be helped to build a latrine for the use of all nearby villagers in the centre of the village, thus increasing the amount of food for the fish and resulting in an increased yield of fish. As I visited projects, the discouraging feature of this idea from my perspective was that the pathway to the latrine was made of single or double bamboo rails, which offered many opportunities for falling in, and the final place for squatting – which, if you were lucky, was protected with a length of sacking – was also precarious. The Javanese, with their tradition of politeness and formality, introduced a new custom for visitors to the village; they were graciously asked, if they felt the need, to unload themselves in the communal fish pond, feed the fishes, increase the stock of fish, and thus increase the yields to the villagers.

What did I learn from the coalface?

First, find enthusiastic and receptive villagers who are prepared to try new ideas and are prepared to put their own energy and enthusiasm to use, not just for their own family but for their own community. This worked well in Central Java, as such people seemed in good supply. There were some corollaries, however: do not assist people beyond their capacity to manage or to take risks beyond what they can afford to lose; and do not introduce villagers into supply chains that are beyond their control. One of the projects that Oxfam undertook with villagers in Central Java was cow fattening – but a cow was a much larger investment than a goat and so much more risky. The cow supply chain also introduced villagers to unscrupulous middlemen who would take advantage of their relative lack of power.

Second, find enthusiastic students who are ready and keen to put their university skills to work to solve the problems of poor people. They may be in technical areas, like the technical school graduates who formed Dian Desa, or lawyers, or agricultural and animal husbandry technicians. In general, it was my finding that if such people took up government jobs, even though they were in their technical area of competence, they found the government ambience of bureaucracy and corruption dispiriting and this was likely to ruin their enthusiasm. The alternative was to encourage such people to form or to work for CSOs and NGOs – which, for the most part, would allow them room to practise their ideas and, indeed, would often suggest to them new ideas and practices.

What happened to it all?

CSOs in Indonesia continue, multiply and progress. There was a period in the latter days of Suharto when the government said NGO (non-government organisation) meant anti-government organisation and all CSOs had to rename and refer to themselves as LSM (*Lembaga Swadaya Masyarakat* – People's Self-help Groups) and had to have government security officers sit in on their meetings. This was soon rescinded after the Suharto era passed. It was also seen to be a waste of time by the security officials long before that. The biggest problem for the future of

CSOs, however, was – and is – their dependence on foreign aid funding. Donors are fickle and short-term, and few people have worked out ways for CSOs to be financially self-reliant, outside of using microfinance and lending programmes. Dian Desa, interestingly, now runs its own shops selling high-quality accoutrements made from sting ray and frog skin, which they have learned how to tan.

Yayasan Dian Desa continues apace, devoting more effort to helping create opportunities for income generation for poor villagers, and being more disciplined in toning down their enthusiasm for AT and applying careful scientific research. This is particularly seen in their work on fuel-efficient stoves, where they very carefully researched calorific improvements around different shapes and dimensions of hand-thrown clay stoves.

Stall-fed goats are now part of the advice of extension workers and agricultural scientists all over the world. Oxfam never claimed to be originators of this, but merely extenders of good ideas received from other practitioners. Excreta feeding fish ponds has been practised for thousands of years in China – we merely helped cross-national learning to take place, and with a little financial incentive allowed poor people to take advantage of such ideas that were new to them.

With the end of the dictatorship of Suharto and the government's suspicion and dislike of CSOs, the opportunities arose for CSOs to develop and demonstrate valuable developmental ideas without being harassed. But while affiliation with the government became less important, political affiliation in the new era of competing political parties became more important, and polarisation because of perceived religious belief became more important as well.

And on a personal note

My job for Oxfam in Central Java was the first one after getting married. My wife, Clare, had long experience of managing a theatre in Glasgow, Scotland, and, in preparation for her to join me in Indonesia, we thought about what relevant field of work this might lead to in international development. We reflected that managing a theatre involved managing a number of small enterprises (scene painting, ticket sales, cafeteria, bar, carpentry) and that these skills could be offered to local organisations working on small enterprise development. Indeed, this worked out well and she was employed by such a local organisation, which enabled her to use her skills.

The Oxfam programme concentrated on drinking water, and this was filmed by the BBC TV children's programme *Blue Peter*, which covered our work in 1981. While there were many springs in upland Java, getting drinking water in the dry areas of Jogjakarta was difficult. The limestone formations had underground rivers that flowed into the Southern Sea (and often surfaced some distance from land in the salt ocean), but access to them was only through very narrow and difficult chasms (made worse when carrying water cans). And in the traditional religions of this area, where any link with the Southern Sea suggested links to the Goddess of the Southern Sea and her consort, the Sultan of Jogjakarta, getting water from underground sources was a sacred task that required the correct attitude in the collector.

I also knew Ann Dunham Soetoro, the mother of Barack Obama, whose father was a Kenyan. She subsequently married an Indonesian and did her PhD research among blacksmiths in this same area of Jogjakarta. She died before she finished her PhD but it was subsequently finished by her friends. I knew her well when she lived in Semarang, as Clare and I did, and when she moved to Jakarta her house was a home from home for us when we visited the big city. I even met Barack Obama as a teenager when he visited her. She was a brilliant development worker specialising in credit and small businesses, fluent in Indonesian and pretty fluent in Javanese, which is much more difficult.[1]

At this time, Oxfam hired field staff on four-year contracts, with a home visit after two years which included an obligation to educate and stimulate Oxfam workers in the UK with stories from the field, and because of Clare's involvement with Glasgow, we decided to go there. Clare and I were concerned about this: we were interested in telling the truth about what we were doing – neither "My trip to the Holy Land with lantern slides" nor showing off about our selflessness. In telling our stories of fish ponds and defecation, there was no problem – "Ach, you mean like the River Clyde" was the immediate response.

As well as our educating Oxfam UK staff on a visit to that country, Oxfam also agreed that the UK staff were entitled to a visit to the field if they had put in a number of years' service. So I received eight Brits to be shown what Oxfam was doing in Central Java. These were people who were interested in development, but actually ran Oxfam shops in towns in the UK and had no experience abroad. They were fascinated by the details of everyday life of Javanese families – their cooking, their child-care, their gardening. A larger understanding of the causes of poverty didn't interest them so much.

During my five years I had a visit from the British Ambassador who wanted to come and see how Oxfam was working with the Indonesian Government – ammunition for his diplomatic discussions. We agreed that he would come to a particular place in Central Java, Klaten, where we worked with landless labourers and the government worked with Master Farmers. Since the Ambassador was there, all manner of Indonesian government officials in uniform turned up. I quickly found that there was no one on the government side who spoke English, and so they agreed that I should be the translator for the Indonesians. Indeed, I translated their mendacious stories about Master Farmers helping the landless people to the Ambassador, and then added my comments to him in English, noting that it was largely lies. The Ambassador gesticulated, went into contortions, and made all sorts of faces at me to tell me, non-verbally, to stop what I was doing, thinking I would cause a diplomatic incident. I told him quickly in an aside that no one from the Indonesian side spoke English, but, believe me, he was sweating.

What were we thinking at the time?

Oxfam had produced a comprehensive manual for its staff:

Pratt, Brian and Boyden, Jo. 1988. *The Field Directors' Handbook: An Oxfam manual for development workers.* Oxfam Publishing, Oxford, UK.

And this had a tremendous bibliography that took you to further manuals and sourcebooks on development. Much of my thinking at this time came from this book.

Yayasan Dian Desa produced many booklets and research papers to underpin their work, and these are available at www.diandesa.org.

We listened to and learned from many Indonesian CSOs, many of which, however, wrote their best stuff for their donors and did not necessarily circulate it widely. In some cases our experiences were taken up and written up later by others:

Tanner, J. C., Owen, E., Winugroho, M. and Gill, E. M. 1994. *Cut-and-Carry Feeding Systems for Small Ruminants.* DFID, London, UK.

Brown, Janet H. and Prayitno, Budi. 1987. Backyard fish farming in Java, Indonesia. *Community Development Journal*, 22 (3): 237–241. www.jstor.org/stable/44256666

Note

1 Janny Scott of the *New York Times* wrote a biography of Ann Soetoro. It contains in Chapter 4 a description of her fieldwork which has tremendous insights into how to work in villages in Java. I was lucky enough to get this from the horse's mouth at the time, but it is a rich source for anyone wanting to work in development. Scott, Janny. 2012. *A Singular Woman: The untold story of Barack Obama's mother.* Riverhead Books, New York, USA.

8 The far east of Indonesia, 1979–84

Oxfam, famine in East Timor and the amazing growth of *Leucaena leucocephala* in NTT

... In which Oxfam works in the very different far east of Indonesia – first with relief supplies in the war in Indonesia's East Timor, and then with helping to distribute the amazing lamtoro agung *seeds to reforest the region.*

Apart from working in Central Java, during my time in Oxfam I was also hustled by them into preparing a response to the entirely man-made famine in East Timor. With the end of Portuguese colonial rule, in 1978, Indonesia decided it was going to occupy the country in order to avoid the possibility of a neighbouring local Marxist State which would oppose Indonesia. In fact, the appalling cruelty and human rights abuses of the Indonesian occupation stimulated the growth of a Socialist and nationalist rebellion among the East Timorese, resulting in such a problem for Indonesia that they finally gave the country back its independence in 2001, after a terrible scorched earth policy that destroyed nearly all that they had built in 24 years of occupation (more on this in Chapter 14 when I went to work in East Timor after its independence).

In 1979–80, word was getting out to the outside world of disease epidemics and famine in East Timor caused by the destructive policies of the Indonesian Army against the local population. Although the Indonesians made East Timor into a thirty-third province (Timor Timur), it was, in effect, a combat zone and training arena for the Indonesian Army, with very little involvement of the civilian population except as objects of control and terror. Oxfam decided to respond to the famine and health crisis by an airlift of food and medicine and the setting up of emergency health posts. This was to be done, however, in a country that did not admit that there was any crisis in their new province, and it had to be done without any foreigners being allowed in for planning, distribution, management or monitoring. My predecessor, Glen Williams, and I, very diplomatically, enrolled the retired Surgeon General of the Indonesian Army, General

Suroyo, who was now running a Catholic hospital in Central Java, to mount a relief effort in East Timor. He was Catholic, as were all of the suffering people of East Timor, and he agreed to do what he could with Oxfam's help. In a very feudal manner he called up all his ex-staffers, now also retired, to come back and work with him, and he got permission from the Indonesian military to undertake a relief project. Oxfam sent me a very large amount of money – General Suroyo lent me two bodyguards as I carried this money around Jakarta buying supplies, and then he talked the military authorities into allowing a civilian plane with emergency supplies to land in East Timor. I was not allowed to visit for a further 20 years.

I said at the start of this book that I rarely operated at the policy or political level. There was one episode in connection with East Timor that was different. Oxfam phoned me up to say that the UK Foreign Secretary, Lord Carrington, was coming to Jakarta, and I had to get alongside him to make Oxfam's feelings known about the UK selling British planes to the Indonesian Air Force as they were being used in East Timor. I pressured the UK Ambassador to give me half an hour in the embassy to meet him before his "proper" meetings with the Indonesian Government. I prepped myself well and was introduced to Lord Carrington, eager to put, forcefully, Oxfam's position.

Good luck to others who deal in such situations frequently, but I failed miserably. Lord Carrington snowed me, asking after me and my family, assuring me of HMG's interest in the situation, and, I am sure, washing his hands of me as soon as he left the room into discussions of further arms' sales to the Indonesian Army. I do not believe it was a very successful policy intervention, but I did get alongside him and said my piece.

One spin-off from Oxfam's involvement in the East Timor emergency was an appreciation of the difference that Eastern Indonesia presented from fertile Central Java. East Timor was the eastern end of the last island in the long archipelago of islands from Java towards the east – Lombok, Sumba, Sumbawa, Flores and Timor. The further east you went the drier and less hospitable was the terrain, and also the more Catholic and animist; East Timor, following its 400-year colonisation by Portugal, was officially Catholic but included a large population that paid homage to indigenous and animist religious systems. Even further east from Timor was Papua, but that was a completely different environment, climate, flora, fauna, population and economy.

The Indonesian Government's development research community was producing more and more information about the poverty and lack of development in Eastern Indonesia and this was centred on NTB (*Nusa Tenggara Barat*) – the western archipelago comprising Lombok, Sumba, Sumbawa – and NTT (*Nusa Tenggara Timur* – the eastern archipelago) comprising Flores, Timor and the smaller islands of Sabu and Rote. Tim-Tim (Timur Timor or East Timor) was still at this time under the control of the army, and had been since the invasion of 1978, was not the focus of any development efforts of the Government of Indonesia, and was off-limits for any foreign NGOs.

One of the greatest problems in NTT was the lack of rainfall; there was also the deforestation that had taken place of the existing tree coverage. This had resulted in very dry land areas, the drying up of rivers, a lack of forage for animals and a lack of firewood for cooking. Oxfam, like many INGOs, responded to the government's call for attention to be paid to NTT province and established a branch office in Kupang, the capital.

A very unique development intervention had happened just at this time – which was filled with rumours and hearsay. As we heard it, a Catholic priest from Flores had gone to Hawaii for some religious purpose and had been impressed with a tree species indigenous to Hawaii which he thought would be helpful in NTT. It was called in the Philippines giant ipil-ipil, and in Hawaii the giant leucaena. It is one of the mimosa family and known in Indonesia as *lamtoro* or *petai cina*, but the new version from Hawaii was quickly called *lamtoro agung* or giant *lamtoro*. Its Latin name was *Leucaena leucocephala*. It thrived in dry areas, it grew vigorously, and its leaves were nutritious to livestock (with the exception of horses – the rumour was that if horses were fed a diet only of *lamtoro* they lost their hair). It could be pollarded for firewood and made passable charcoal. Because it grew so fast and could be pollarded, using it for charcoal did not produce barren and deforested land. It was also a nitrogen-fixing plant so was a natural fertiliser of barren land.

The rumour was that the priest had brought back seeds of the *lamtoro* in his cassock, had not declared them to the authorities and had set up some seedbeds for them back in NTT. They had done very well, had grown beyond expectations, and were even in some places considered a weed that needed to be controlled.

The local people of NTT were hugely impressed with *lamtoro agung*. It did not need any elaborate planning protocols and fruited with many seeds after 18 months. There was a great desire for the seeds and a number of early innovators set themselves up as seed producers and seed merchants. I remember visiting one well-set-up farmer in Timor whose house had furniture, a TV and a sound system. "*Lamtoro* seeds have produced all this," he said, gesturing to his wealth and a relatively small plot of young trees with the sacks of seeds that he had harvested and was selling on to his neighbours.

Lamtorisasi (Indonesian for the process of spreading *lamtoro*) was a self-perpetuating process. Amazing claims that rivers had come back into flow, that rainfall had improved and that areas which had been brown were now green were repeated everywhere. There were plans for aerial scattering of seeds over the territories of NTT, although this never happened as far as I know. To a large extent this was indigenous development – Oxfam's role, initially, was to organise learning visits between areas that had already planted the miracle trees and those that were late developers. It became accepted that animals would not thrive on 100% *lamtoro*, but that they would certainly improve on mixed grazing of *lamtoro* and grass.

Lamtoro was also a nitrogen-fixing plant with deep roots which brought back fertility to areas where erosion had scoured out the nutrients – and this produced a further use of the *lamtoro*: for contour terracing. If the *lamtoro* was planted densely on the contours, then it would make a small dam to hold back surface water flow, it would catch leaf and soil run-off to build up a fertile area uphill, and it would stop erosion and gullying.

Taking advantage of all these potential benefits of *lamtoro* required some further support work from Oxfam. The principal one was encouraging the making an A-frame level: i.e. three poles in the shape of the letter "A" with a plumb line dangling from the apex and a mark in the centre of the crosspiece. Farmers would carry the A-frame across the countryside, linking places of the same level as shown by the centred string, and then they would plant large numbers of seeds of *lamtoro* along the contour line, before the rainy season. In the rainy season the *lamtoro* would send down roots, anchor a line of plants along the contour, and catch water and vegetation run-off. Once the area behind the contour terracing had become established and fertilised by the nitrogen-fixing aspect of the *lamtoro*, farmers could plant a range of crops (maize, peppers, cassava, etc.) along the contours.

A final beneficial aspect of *lamtoro* comes when the leaves are cut and pruned and laid on the soil around the terracing. The *lamtoro* vegetation is a green manure and helpful to maintain soil fertility.

Oxfam worked again with Yayasan Dian Desa, which also opened up an office in NTT, to encourage the use of *lamtoro agung*, to manufacture and use the A-frames, to set up contour terraces across the land, and to teach many simple agricultural techniques to drive benefit from the amazing *lamtoro agung*.

What did I learn from the coalface?

Keep your ear to the ground about local innovations. What was most impressive about the benefits of *lamtoro* was that it could not be monopolised or held back by some to the exclusion of others. Everyone could benefit without barriers to the use of the technology. One of the advantages of being an INGO, like Oxfam, was that it was able to subsidise its representative and core staff to search out and appreciate new ideas and new organisations. This in turn would lead to new programme possibilities. Another advantage was that, once the NGO had identified a good local CSO, it could keep supporting it, through the vagaries of donor funding, and invest in its institutional growth, helping with training, communications and organisational structures. Oxfam was good at this and has received many tributes from its local partner organisations that it stayed with them and helped them to grow.

It was also important to be able to put yourself in the shoes of the local government, which was invariably suspicious of the role of foreign organisations that claimed that their aid would help the country's poor. The most extreme version of this was having to explain INGO politics to the Indonesian Government. Oxfam, at that time, had an Australian affiliate called CAA (Community Aid Abroad), which had taken a much more confrontational role vis-à-vis Indonesia (not surprisingly, since the Indonesian Army killed Australian journalists there). I had some delicate balancing to do to let the Indonesian authorities know that Oxfam UK was different from CAA, and that we were working through their retired Surgeon General of the Indonesian Army.

What were we thinking at the time?

News from East Timor was difficult to access and often came with a strong political spin from the political activist organisations in Australia, Ireland, UK, USA and different countries of Europe which were vehemently lobbying for East Timorese independence with their governments and the UN. What was clear, however, was that terrible things were being done to the population of East Timor, and this was subsequently revealed in published research.

van Klinken, Gerry. 2012. *Death by Deprivation in East Timor 1975–1980.* World Peace Foundation, Cambridge MA, USA.

Mounting a large relief effort in a country where Oxfam was not allowed to monitor was a gamble, but our trust in retired Surgeon General Suroyo and his co-retirees was, as far as we could find out, vindicated.

In respect of *lamtoro agung*, we were involved at the time in a breaking story in NTT province, and this was subsequently written up and became scientific orthodoxy.

Wirjodarmodjo, Hartono. 1983. *Leucaena Leucocephala: The Indonesian experience*. FAO, Rome, Italy.

Tacio, Henrylito D. 1993. Sloping agricultural land technology (SALT): how to farm hilly land without losing soil. Technical note 72. FAO, Rome, Italy.

9 Positive deviance, 1980–81 and 1984–85

Nutrition in Indonesia and rice/fish farming in north-east Thailand

... In which positive deviance is documented – local people who found their own ways to solve poverty and malnutrition: in Jogjakarta, Indonesia, through collecting and feeding children the winged bean; and in Thailand through growing both rice and fish in the same paddy field.

While acting as the Country Director of Oxfam in Jogjakarta, Central Java, I supported the work of a local Protestant church organisation called CD Bethesda. It was the community development unit of the Jogjakarta-based Bethesda Hospital. They concentrated on improving malnutrition in the dry, poor and malnourished area of Jogjakarta. Later on, after I left Indonesia, I worked as the Asia Director of the Canadian organisation CUSO – whose main work was placing Canadian volunteers (or *coopérant* in French Canadian) in useful developmental positions. In Asia, CUSO worked in Bangladesh, India, Thailand, Indonesia, Malaysia, Sri Lanka and the Philippines, but this example of its work comes from how its *coopérants* in north-east Thailand, working with local Thai organisations, concentrated on encouraging raising fish in rice paddies. Both organisations worked with individuals who had the same constraints as everyone else yet found a way to succeed. They illustrate positive deviance – letting go of traditional ideas about authority and power and instead allowing people to discover answers for themselves.

Positive deviance is a truly revolutionary concept that is founded on the premise that at least one person in a community, working with the same resources as everyone else, has found a way to overcome a problem that continues to confound others. This individual, or group of individuals, is an outlier in a statistical sense (i.e. a deviant) – an exception, someone whose practices deviate in a positive way from the norm. He or she (or they) may have discovered something themselves, may have picked up on someone else's idea and decided to apply it, or may be doing something that they are unaware of but that has very positive results. It was first conceptualised in the field of nutrition in Vietnam by Save the Children with Jerry and Monique Sternin, and gradually permeated the thinking of grassroots development workers.

I can claim no responsibility for introducing the thinking behind positive deviance to the people with whom I worked – but I encouraged it and helped it to spread. Dr Paulus Santosa of CD Bethesda had had contact with Jerry Sternin and was able to apply his thinking to malnutrition in Jogjakarta. The CUSO volunteers in north-east Thailand and the Thai organisations they worked with were simply smart observers of what innovative farmers were doing in the region; they got behind them, encouraged them, and encouraged the dissemination of their ideas.

The problem with seeking out and applying the positive deviance approach stems from the attitude and mindset of the "expert". Experts think that they know what needs to be done, that they know more than those "below" them, and that change is better driven from the top than from the bottom. Those of us who are conditioned to think like an expert do not start by seeking knowledge from the grassroots and may well miss very important innovations at that level. The two examples given in this chapter illustrate people and organisations with which I was lucky enough to be associated, that listened to the grassroots, and that helped along those who had found or discovered unique solutions to problems that were felt to be insoluble.

Jogjakarta and malnutrition

Jogjakarta is a special region of Central Java and is given a unique status in Indonesia because it has a very old and traditional sultanate, and because the Sultan fought valiantly to defeat the Dutch at the time of the War of Independence. Geographically it stretches from the fertile volcanic slopes of Central Java to the dry and infertile limestone hills that end in the south coast and the sea. For those interested, it is also the area where Ann Dunham Soetoro, the mother of Barack Obama, did anthropological research into blacksmiths for her PhD at the University of Hawaii.

While the city of Jogjakarta is a vibrant and productive centre, the rural areas of the district have some of the worst nutrition data in Indonesia. There are very few rice fields and limited fertile land, and, as a result, limited income. Local people make a lot of craft items from local grasses and leaves, but these sell for very small amounts in the city. Children have high levels of stunting and chronic malnutrition in general is widespread.

At this time, the Protestant CD Bethesda Hospital in Jogjakarta city opened its "Hospital without Walls" policy and undertook nutrition surveys in the rural hinterland of Jogjakarta.

One of the ways of surveying nutrition of under-fives in Indonesia at the time came from UNICEF, which introduced baby weighing at monthly meetings of local "*arisan*". *Arisan* were traditional monthly savings and credit groups, especially in Java, and nearly always of women. Women came together once a month to make small contributions and for one of the group in turn to collect the kitty. UNICEF noted that when women came together for such meetings they usually brought their babies and young children with them. Working with the Ministry of Health, UNICEF started to introduce systematic baby weighing at these meetings. Baby weighing was linked to monitoring of the babies' weight over time and advice about what to do when the baby's weight was subnormal and continued to be so. CD Bethesda took up this innovation and soon found that the majority of babies in the rural areas of Jogjakarta were below normal weight for their age – i.e. they were malnourished. Dr Paulus, however, went one step further – he noticed that malnutrition was not spread equally among the children of women attending the *arisan* baby weighing meetings. A few of the women's children seemed adequately nourished, and he wondered why this was happening.

He pursued this, talking to the mothers of adequately nourished children, and tried to find: (1) whether their situation was different in any way; and (2) whether they behaved differently to their children in any way. He found to his surprise that the women with adequately nourished babies were no different in any way from the rest of the women attending the meetings. This was an important conundrum. It seemed counterintuitive. These women's children were doing better than their peers, but there was no clear reason for this. Dr Paulus and his staff talked further to them, observed their behaviour, and, above all, sought to persuade them that he was truly interested in learning from them, not in isolating them or harming them in any way.

And slowly it became clear that the women of the better-nourished children were assiduous collectors of wild plants, which they added to the small amount of rice that they gave their children. To a certain extent all women collected wild plants to add to their children's food, but the mothers of better-nourished children were, it seems, more assiduous in collecting such food and feeding it to their children – and they collected in particular one plant called the "winged bean". This plant is a twining annual vine, called *kecipir* in Indonesian, and is very high in protein. It is found in the wild and also cultivated in some home gardens. It soon became obvious to Dr Paulus and his staff that there were a few women who collected and fed a greater variety of wild plants, including *kecipir*, to their children, and these mothers' children were healthier and better nourished than those who did not. These mothers were "positive deviants": i.e. they were different from the other women in their behaviour, but different in a very positive direction.

Once the link was made between the behaviour of these women and the better weight for age of their children, as revealed in the monthly weighing sessions at the *arisan*, it was the task of CD Bethesda to persuade the majority of mothers to change their feeding practices. This was not a simple task, as people in general (including Javanese mothers) are very conservative about diets and eating habits.[1] The approach of most development workers is to work from a basis of "expertise" – those at the top know more than those at the bottom – and it is assumed that when the "advice" of the experts is proffered, those at the bottom will accept it and will roll it out through the relevant people. To say, in effect, that your peers have the answer, rather than experts, and you need to learn from them is, for many, counterintuitive. It takes time. It is, however, very relevant in situations where top-down suggestions (which often involve outside resources) have not worked. It depends on what resources are available to the peers but does not require anything outside the resources of the local community.

My role in all this was very different from what I had done before (in Ethiopia, Sudan and Botswana), where I had been hands-on, and the active promoter, practitioner and manager of different suggestions and innovations. The strategy of Oxfam at this time (and of the best of the other NGO donor agencies) was – and still is – as a catalyst: to look for and find competent, clever and well-motivated local organisations, and to get behind them to support them in their work. Some of this was personal support – you supported someone who had very good ideas but could not afford to leave their day job to implement them; some of it was supporting the growth and development of a band of people who wanted to work together to do something useful but were finding it difficult – helping them get access to information, to meet others of a similar attitude and experience; and some of this was helping a group of enthusiastic friends make the transition to a local organisation – from volunteers to paid staff, from part-time to full-time practitioners.

A further form of help, which was more complicated, was to endorse and promote their work to local and national government and to larger donors as their requirements started to exceed what Oxfam could do for them. Oxfam's advisory document for field staff, *The Field Directors' Handbook*, suggests the following non-financial assistance to groups and local organisations:

- inter-project learning – putting groups working on the same topics in contact with each other; discussions – giving them the opportunity to try out their ideas on someone with a trained, critical and experienced mind;
- participation – assisting technically qualified project planners to be more aware of the social, cultural and other needs of the group; mobilisation of local experts – particularly those who can provide sustained support/advice;
- information and publications – including making grants for groups to acquire valuable literature, films, videos, etc.

I became more of a mentor and fellow enthusiast for the work of local Indonesian organisations and started to become embroiled in the complicated politics of NGOs. At this time local bodies were usually called NGOs (non-government organisations), and to a suspicious government they were often seen as anti-government organisations. Many governments, including that of Indonesia at this time, took the perspective that they alone were the source of all good ideas, by right, and that if anyone had a different perspective, it was likely that they were on the wrong track (because government officials should be considered the experts) and, more worryingly, they were trying to win the people's loyalty away from the government. Once I, as Field Director of Oxfam, had discovered CD Bethesda and had satisfied myself that it was an organisation of integrity, competent implementation and good management, then it became a question of developing close relations with those who ran CD Bethesda and helping them with problems as they came up.

It also required the donor organisation – in this case, Oxfam – to be suitably humble and prepared to learn from the local organisation, which had much more grassroots experience than Oxfam. It was CD Bethesda that taught me about positive deviance, not the other way round. It was my job, once convinced of the value of their approach, to help them to run with the idea and develop it further.

Linked to all these ideas of mentoring and personal support to the organisation was the very basic relationship of funding. CD Bethesda (and indeed any other Indonesian NGO with which we worked) was most interested in being funded to do the work that they had identified and wanted to do. It was important to establish a professional and mutually satisfactory relationship for the funding so that it did not deteriorate into one of begging on the one side and handouts on the other. Helping to create a budget, helping to make sure that funds were allocated as agreed and helping them to write financial reports that would satisfy auditors were all important parts of the relationship.

My pleasure in this kind of work was gained from visiting the field with the local organisation and seeing what local people were actually doing. Looking over Dr Paulus's shoulder as he talked with local women about choosing wild plants for their children, and then observing the chaotic monthly weighing sessions as mothers showed off their children's growth (or sadly noted their lack of growth). There was, however, the chance that the occasional visit of a foreigner would disturb the dynamics of the local organisation's relations with its clients, and this had to be watched carefully.

The concept of positive deviance as a development approach has been taken up in many different places and in many different ways. The seminal book is *The Power of Positive Deviance: How unlikely innovators solve the world's toughest problems*, and the knowledge of the different ways in which it can be used are accessible in the International Positive Deviance Collaborative. Enthusiasts have tried the concept out in management, industry, education and many more fields.

North-east Thailand and rice/fish cultivation

North-east Thailand is dry and infertile compared with the lush and productive rice fields of the central plains of Thailand. It is called Isarn locally, which translates as "weeping prairie" – a comment on the starvation that has been associated with Isarn for a long time. Further evidence of its poverty is its cuisine; it contains many insects, bugs and lizards that no one else will eat. It has unpredictable rainfall and quite frequent floods. The standard of living and the level of wages are the lowest in Thailand, as a result of which many people from Isarn migrate to Bangkok, where they work on building sites, as domestic workers or as prostitutes to send money back home.

I was living in Penang now, with wife and two children, and working for CUSO, the Canadian volunteer (or *coopérant*) placement agency. It was most active in Thailand, providing *coopérants* in teaching and disability, but also in agriculture. I went there many times – particularly to Isarn, which abuts Cambodia – where CUSO was working on what was called "alternative farming". It was called "alternative" because Isarn could not use the high-input, high-output farming system of the rest of the country which depended on highly fertile soil, intensive irrigation and the use of the Green Revolution package of imported seed, heavy fertiliser and heavy pesticide use. Gerry Van Koeverden, the CUSO agricultural coordinator, could not see such systems fitting the resource base of Isarn, and so became very interested in seeing what agricultural innovations local farming pioneers were employing. And the most notable one was getting two products off one piece of land – that is, rice and fish – with the fish raised in the rain-flooded rice paddy. The fish were caught in the gullies between the rows of rice and provided extra protein for the families growing rice, or were caught at intervals and sold.

Down in the central plains, raising fish in rice paddies had been common in the past, and the use of year-round irrigation made this relatively easy, but because of the use of pesticides associated with modern farming, the fish started dying off from 1970 onwards. Farmers either separated their rice farming from their fish farming, or stopped breeding fish altogether. In Isarn, however, fish were raised in rain-fed rice fields and were helped because the local Isarn farmers had not bought into the full Green Revolution package of imported seeds, pesticides and fertilisers and so could still grow the two together. The government body that had helped with rice/fish cultivation (the Centre for Rice–Fish Culture Research) was started in 1968 and had addressed the farmers of the central plains, but it had closed down in 1974 – and thus was not, unfortunately, available for servicing the needs of the Isarn farmers.

In Isarn, farmers caught fingerlings in local rivers or streams, put them into the fields at the time they flooded the rice shoots, and raised them along with the rice. The farmers dug pits at the edge of the rice fields, and, as the water level fell, the water and fish drained into the ponds where they could be kept alive for many months and netted out for food or sale.

It was not just fish in rice fields – the dynamics were much more interesting. The fish ate lice and small insects around the roots of the rice stems, performing an insecticide function, and farmers could throw buffalo dung or trash from the fields into the rice paddies to feed the fish. The rice/fish farmers in Isarn were initially (in 1987) positive deviants, although their success soon encouraged others to join that system of farming. What was unique was that this was occurring spontaneously among farmers – not being pushed by any government development programme. As Janet Durno, herself a CUSO *coopérant* at this time, wrote: "The true impetus for the development and spread of rice/fish farming lies with the farmers themselves."[2] CUSO, working with Thai

NGOs (the Farming Systems Research Institute and the Appropriate Technology Association), was encouraging them.

CUSO's involvement (and thus my involvement, since at this time I was the Asia Director of CUSO) was to bring Canadians with a background in farming to Thailand (and more particularly to Isarn): in effect, to de-educate them about the farming knowledge that they had from their Canadian heritage and to re-educate them to observe very closely (through on-farm research) what the Isarn farmers were doing, and be as supportive as they could, helping them to collect data and undertake experiments – particularly to ascertain the contribution that fish made to pest control.

I would accompany the CUSO *coopérants* as they carried out their weekly interviews with rice/fish farmers. The volunteers were given training in Thai, but there was a very different dialect in Isarn, which was closer to the Lao language. It was fascinating to watch them holding detailed discussions with farmers up to their knees in the rice paddies, seeking the opinions of farmers about this new technique. The job of CUSO Thailand was to make sure that the volunteers were not leading the discussion, or being seen as the source of received wisdom.

For me, this was only one of the programmes of CUSO in Thailand, and I was responsible for managing many programmes of volunteers in Malaysia, Indonesia and Bangladesh, as well as Thailand, but in no other place did the volunteers have the chance to listen to positive deviance being played out in their work.

What did I learn from the coalface?

I learned the importance of watching and listening to what innovative local people were doing to improve the lives of themselves and their children, and the importance of supporting the development of organisations that had bought into this idea and were practising it.

The other important learning was that the conventional wisdom about agricultural innovation – i.e. the Green Revolution – had limitations in specific localities. While there is consensus about how the Green Revolution has increased yields, there are important variations where this does not take place, or where it takes place at the expense of other locally important benefits.

The next complication was persuading the local Ministry of Agriculture, and the government of a country that had bought into the Green Revolution, that local variations were important and worth their attention.

What happened to it all?

Indonesia

Looking back 20 years later, the practice of scavenging for nutritious wild plants, particularly the *kecipir* or winged bean, was taken up by UNICEF, by the Ministry of Health and by PKK (the welfare organisation of civil servants' wives). It has become standard practice and education about it is a regular part of the *arisan* monthly meetings, along with their baby weighing.

Thailand

Unfortunately, in 2019 there are very few examples of rice/fish cultivation continuing in Isarn. On a return visit I was told the same story as had happened in central Thailand. More and more farmers, under pressure from the government's Ministry of Agriculture, which is in turn under pressure from the major agricultural suppliers, have used amounts of insecticide that are poisonous to fish, bringing to an end this mutually supportive farming system. CUSO volunteers subsequently took their knowledge and enthusiasm to Laos, where I believe it is still being practised.

What were we thinking at the time?

Development practitioners at this time were learning from each other, with occasional informal papers. Later, the fields became more formalised and more accessible through books and websites. The site for positive deviance is www.positivedeviance.org and a new International Positive Deviance Collective was launched in 2017 and gives interested practitioners information about the steps involved in the positive deviance approach.

The core reference book (not available to us at the time) is:

Pascale, Richard, Sternin, Jerry and Sternin, Monique. 2010. *The Power of Positive Deviance: How unlikely innovators solve the world's toughest problems*. Harvard Business Review Press, Harvard MA, USA.

Similarly, information on winged beans and rice/fish was circulated informally and then established more solidly, sometimes through conference papers:

MacKay, Kenneth T., Chapman, Greg, Sollows, John and Thongpan, Niran. 1986. Rice-fish culture in north east Thailand: stability and sustainability. Paper presented at the IFOAM 6th International Scientific Conference, University of California, Santa Cruz, USA.

National Research Council. 2011. *The Winged Bean: A high-protein crop for the tropics*. National Academies Press, Washington DC, USA.

The work of Dr Paulus and CD Bethesda was noted by Robert Chambers:

Chambers, Robert. 1983. *Rural Development: Putting the last first*. Longman, New York, USA, p. 202.

Janet Durno's article on rice/fish practices in Thailand is:

Durno, Janet. 1987. From rice/fish to integrated farming: sustainable agriculture for resource-poor farmers in north-east Thailand. In Holloway, Richard (ed.), *Doing Development: Government, NGOs, and the rural poor in Asia*. Earthscan, London, UK.

And, as before, the key manual for development practitioners is:

Pratt, Brian and Boyden, Jo. 1988. *The Field Directors' Handbook: An Oxfam manual for development workers*. Oxfam Publishing, Oxford, UK.

Notes

1 Sternin's book *The Power of Positive Deviance* points out a cross-cultural version of this: "The idea of doing so [introducing new foods] seemed as far-fetched to many as feeding garden snails (enjoyed by the French as escargots) and dandelion leaves (which garnish signature salads) to the children of Westerners."

2 See Chapter 9 in *Doing Development: Government, NGOs, and the rural poor in Asia*.

10 Bangladesh, 1989–95

NGOs, CSOs, dependence on aid and independence from aid

... In which Bangladeshis come to understand the third sector in Bangladesh, as organisations moved from defining themselves by what they were not (non-government) to what they were (civil society), and discovered the complexities of fighting government for their place in society.

Arriving in Dhaka, Bangladesh, one was assailed by the density of the population – roads full of people pushing carts, pedalling rickshaws, driving motorcycles and tuk-tuks, and also driving larger and ramshackle lorries, all interspersed with mothers, fathers, many children, many old people, many beggars, and those who wanted to sell you things. Even when you left the city the density of the population was mind-boggling, denser than Hong Kong, denser than India, and all made possible by the annual floods of the huge rivers that deposited fertile silt on the fields, allowing people to grow food and cash crops. If you travelled into the country, you would soon come across the gigantic rivers so wide that you could not see the other side, and a new density of river craft – full-sailed boats carrying rice, smaller engine-powered barges, and large canoes carrying people, bicycles, goats and all manner of goods. The rivers, when they flooded, could flex their mighty muscles and break off a few acres of riverside land, tipping it, and possibly whole villages, into the river.

This sounds like a country with great potential, and this is true, but the lives of Bangladeshis were generally very poor, kept in poverty by a very hierarchical social structure, with huge numbers of people landless and in debt to very rapacious landowners and money lenders. It was (and is) also a violent country, with many thugs employed by richer people to enforce their wishes on poorer people. Violence against women was and is widely accepted throughout the country. However – and this is the redeeming quality of Bangladesh – it also has very strong and empowered people who form organisations to defend themselves. Bangladeshis look back with grief but also with pride to the War of Independence in 1971 when they broke free from Pakistan and established their own country with its own language, and its own organisations.

My family and I spent six years in Bangladesh recruited by a US NGO called Pact, which was operating a USAID-funded project (PRIP)[1] for the development of the capacity of local citizens' organisations. My time in Bangladesh educated me greatly about the nature of citizens' organisations, particularly in an unfriendly governmental environment. But Bangladesh had plenty of excitement to offer as well: during our six years, one flood, one cyclone, one revolutionary regime change which almost led to civil strife between the NGOs in the country, and, from my perspective, the evolution of my successful foreign-funded project into an indigenous NGO (the PRIP Trust), which has continued to the present, and the evolution of my deputy, Aroma Dutta, into the manager of the PRIP Trust and now an MP and a member of the Human Rights Commission of Bangladesh.

Not many people know that Bangladesh boasts the two largest NGOs in the world: BRAC (founded by the late Sir Fazle Abed) and the Grameen Bank (founded by the Noble Laureate Muhammad Yunus). Both of these organisations have had huge influence on the growth of CSOs and NGOs around the world. For anyone interested in learning about civil society organisations and the possibilities and constraints of civil society in a hostile governmental environment, Bangladesh is the place to serve your apprenticeship. My six years also covered the period that gave birth to the term "civil society organisation" (CSO) at the first World Environment Day in 1992, with all the rethinking and analysis this entailed. It was also the start of the increasing amount of support that international donors were putting into the work of such organisations, with all the questions that they were asking about what they could hope to get from this support, what capacities such organisations should have, and how those capacities could be further developed. Increasingly heated discussions derived from whether CSO staff should be paid, and how much; what level of equipment and supplies they ought to enjoy; whether they could get involved in business; where their funding should come from; and the extent to which they should be involved in politics and local government.

Bangladesh had huge numbers of very small citizens' organisations, and yet, at the same time, the largest CSOs in the world. It soon became clear that my work was going to address capacity building of the smaller Bangladeshi CSOs, while learning from the larger ones.

My arrival in the country coincided with the extraordinary floods of 1989, and I was straight into hiring a boat, travelling out to one of the biggest NGOs (Gonoshasthaya Kendra), which had been marooned by flood water outside Dhaka, and ready to commit to a programme of food relief for very poor people living in their vicinity on the roofs of their houses with water up to the eaves. I came in to the airport, relieved to find some extant concrete for the runway, since water lapped around its perimeter. Also on the plane was a film crew from CBC News (Canada) who were very interested in filming human interest stories. I traded local knowledge for coverage of the work of the organisations with whom we were working, and we all jumped into a motor-powered boat so that the interviewer could hold shouting conversations through a translator with poor people on their rooftops.

Bangladeshis are used to floods every year, but occasionally they are massive – which was the case in 1989. After the floods had gone down, I mounted an interesting project to try to prepare villagers for the next time there was a massive flood. I sent a village worker plus a Bangladeshi artist to talk to communities who had lived through the flood and to get their stories of what they remembered of the floods, and particularly what innovations they had adopted that were worth remembering for next time. The artist drew their stories and checked with them that he was representing their memories. Some of these were: keep a 20-foot length of piping to add to the height of a handpump so you can keep pumping clean water even when the pump itself is covered in flood water; build fences around your fish ponds from mosquito netting so that fish cannot escape from your ponds when the floods rise above the embankments of the ponds; plant seeds in seed beds made of rafts of floating banana stems so that germination is much faster as soon as the floods have gone down. All innovative ideas from the people affected – no outside intervention.

The flood also introduced me quickly to the complex dynamic of donors, CSOs and people in Bangladesh. At the time of the floods all Bangladeshis were either protecting themselves and their neighbours, or, if they belonged to a group, were trying to find a donor who would give them the money to mount a relief operation. At the same time, donors were keen to be seen to be active in helping those whose lives were harmed by the flood water, but were aware that they were simply seen as cash cows. It was very easy for it all to degenerate into self-congratulatory heroes on the one hand and passive begging victims on the other. The reality was that most people managed their own relief and rehabilitation, albeit with difficulty. Donors were increasingly wary of enthusiastic small CSOs that wanted money to "help the victims" but may not have had very strategic thinking. Increasingly, donors were requiring competent surveys of numbers affected, ways in which they had been effected, and prior acquaintance with the communities of those affected. Some CSOs were practised and professional at times of natural disaster, but there were plenty of others that jumped, without any qualms, from a programme of handicrafts or banking into one of flood relief – long on enthusiasm, but a bit short on experience. All too frequently the cry was for donor funding – with the claim that without that nothing could be done.

Once the floods had gone down and life for poor people had returned from extreme distress to its usual level of distress, I got to work on the main task of "capacity building of NGOs". It came at an important moment for development in Bangladesh. Bangladesh was very poor[2] and very dependent on foreign donors for funds. Many donors were enthusiastic to help Bangladesh but were apprehensive about providing support to the government, which was largely seen to be incompetent and corrupt. The alternative, for many donors, was to support non-government organisations – the NGOs so frequently mentioned. Donors, however, had a very restricted vision of what NGOs were and what they could be. They saw them very much in terms of the voluntary organisations with which they were familiar in their own countries (and since I was working for a USAID-funded operation, we used USA-coloured spectacles). The name of the game we found ourselves playing, therefore, was to train up Bangladeshi CSOs so that they could receive foreign donor money on the terms that the donors were prepared to offer. Was this supporting CSOs or just turning local NGOs into USAID contractors? A lot of very hot air was expended on this topic.

In order to receive USAID funding, local NGOs had to sign a contract (which was not surprising), but the USAID contracts had an astonishing list of conditions to go with them. USAID obviously wanted to protect itself from fraud, corruption and issues that would be injurious to the USA, but they had not been smart enough to see that the conditions appropriate for arms suppliers, for example, were different from those appropriate for local welfare organisations. I am sure that the USAID bureaucrats did not expect anyone to read their voluminous lists of conditions; my favourite bizarre condition was that an organisation taking USAID funds "could not use the funds for artificial weather modification" – not an activity likely for Bangladeshi NGOs. In order to embarrass the Americans and show how pointless was their over-zealous legalese, I got them all translated into Bangla so that local organisations could see how the donor thought about local CSOs. There was no getting round it, however – local CSOs had to sign to swear that they would not get involved in weather modification.

More pragmatically, CSOs and donors together appreciated that NGOs that were serious had to acquire capacities to do what they had determined needed to be done, and capacities to be able to do this sustainably, if they were going to receive foreigners' money. A lot of work was involved in clarifying the optimum level of practices of governance, participation, human resources, management practices, planning, financial management, external relations (especially relations with government) and, most importantly, delivery of the services for which the CSO was established. Many of the NGOs said, strongly, and with justification, to the donors: "You are trying to make us like the government, and our value lies in the fact that we are not government." What, therefore, was a mature and professional CSO? Much of my work in Bangladesh involved trying to get this clear and then following up on what was agreed to help them reach this level.

Bangladeshis are nothing if not argumentative and tend to hold extreme positions and make emotional accusations. Accusations shot around the NGO community, as in many others: should such and such an organisation have a new Toyota Land Cruiser for its work? Should it go to expensive hotels for its workshops? Should meetings take place sitting on the floor (as they did in India) or should they use desks, computers and PowerPoints? Did it all depend on what the donors allowed you to do, or should CSOs have their own standards? The wrangling went to even greater lengths – should all CSOs have members, or only clients? Should CSOs contest local government elections? Should CSOs spend their time trying to get government services, or (realising how bad they were) should they try to make their own parallel version of government services? What should CSO workers get paid?

These were heady times: PRIP's mission evolved into capacity building of local CSOs. With local CSOs, it slowly catalysed a picture of a mature NGO: we gradually agreed the relevant components, we slowly worked out how to monitor the extent to which the NGO achieved these components, and then we designed all sorts of training courses to help NGOs reach the required standards. What evolved, and has been used all over the world, was the OCA (Organisational Capacity Assessment).[3] See Figure 10.1, p. 98.

And behind it all was the government. The President, Ershad, leery of the growth of NGOs (which he saw as a threat to his popularity as well as his country's government's income), demanded a map of the NGOs in Bangladesh. When this was presented to him with the NGOs marked on it, he contemptuously threw it aside, saying that it looked as if Bangladesh had chicken pox. I visited a Permanent Secretary once in my line of work who angrily refused to accept that any NGO staff person should receive a salary – they all had to be complete volunteers, like priests or imams. He was a person who, once he had finished with a file of papers, would simply throw it on the floor for his office workers, significantly called "peons", to gather it up and file it. He was not unique.

In the middle of all this came the Earth Summit in Rio de Janeiro of 1992. Qazi Faruque Ahmed of Proshika Kendra, a large Bangladeshi organisation that PRIP worked with, and who attended it, said to me: "I went to Rio as someone from a non-government organisation [NGO], and I returned as a member of a civil society organisation [CSO]." The terminology changed – people were not defining themselves by what they were *not* (i.e. non-government organisation), but by what they *were* (i.e. civil society or the third sector). The NGOs (now generally agreed to be CSOs) were very impressed by the terms and the narrative of the ex-Soviet Union countries – especially Solidarność from Poland.

It became all the more important for Bangladeshi CSOs to define what they were on their own terms, not just as contractors for foreign aid organisations. However, the jealousy of the small organisations was palpable as they saw the larger NGOs/CSOs pulling down larger and larger amounts of money from the donors. There was almost the feeling that you had to know the right buzzwords in order to unlock the donor bounty – particularly in five-year chunks. The sharper NGOs/CSOs were aware of the fact that donor bounty came with its own limitations – it fitted donor administrative convenience, it had to be defended in donor country parliaments, it was very dependent on donor "fashions" and it was rarely sustainable – but many smaller CSOs saw it simply as undifferentiated and highly desirable funding.

In the midst of all of this philosophical and ideological balancing came regime change. The government of President Ershad fell to general enthusiasm. Ershad had, however, through his networking, politicking and provision of favours, found supporters among some NGOs/CSOs. The coordinating body for NGOs/CSOs in Bangladesh was called ADAB (Association of Development Agencies in Bangladesh) and the time came for an election of office-bearers around this time. The CSOs reflected the politics of the country, and there were very different factions making their cases for office-bearers. The organisation I directed, PRIP, provided services of training and mentoring to many Bangladeshi NGOs/CSOs and was heavily involved in lobbying for competent and professional NGO/CSO trainers and advisers. When the time came for the meeting I was astonished to find that all who attended were searched for weapons, such was ADAB's concern that violence might break out. It did not break out, but some serious schisms were created that have continued for many years.

In spite of my difficulties with USAID, I was very surprised to be approached in 2003 by the USAID official who managed the funding to Pact, and thus, by implication, the services to the NGOs/CSOs that Pact/PRIP provided. He said:

> Pact has done valuable work for the CSO sector, but it should not come to an end simply because USAID's five-year grant to Pact comes to an end. We would like to help Pact midwife and nurture a Bangladeshi-owned organisation to carry on the good work, so that it is not just a foreign-funded time-limited project.

This was music to my ears, not just because it was a confirmation of the value they gave to our work, but because we were talking about breaking out of donor control and administrative limitations to produce something wanted and governed by Bangladeshis. We agreed that my deputy at PRIP, Aroma Dutta, would take over the management of the new organisation, we established it legally as a "trust" and we found a stellar board to govern it, including Sir Fazle Abed of BRAC. We were making sure that the large NGOs would be paying attention to the needs of the smaller CSOs.

To my surprise I found that this idea of evolving from a foreign-funded project to an indigenous development organisation was very rare in development circles, even though there must have been occasions when it seemed a sensible route to follow. We employed a very well-known NGO from India, PRIA (Participatory Research in Asia), to mentor the evolution, and we wrote a manual on how to do it for Pact called *Exit Strategies: Transitioning from international to local NGO leadership*.

Then in 1994 a massive cyclone hit the south of Bangladesh, particularly towards the Burmese border. In spite of all the training and mentoring that PRIP had been doing for five years, most CSOs dropped all their development work and rushed back to being relief organisations. This was, I think, for two reasons: because their countrymen were suffering and they wanted to help them and to show what they could do; and because there was money available for relief and rehabilitation. After the recovery I repeated the exercise from 1989 and the floods – supplying an artist to ask them what happened and what they would do better next time. The lessons to be learned were to prepare to move to cyclone shelters early, to prepare your boat (if you had one) and tether it to make sure it could not get away, to tie your house down and to get your cattle to high ground as soon as the warnings were heard.

During the previous years I had brought the Asia Pacific Disaster Preparedness Centre from Bangkok to teach in Dhaka, and they had promoted Harvard University's *Rising from the Ashes: Development strategies in times of disaster*, which promoted analysing communities for their capacities and vulnerabilities – not just looking at people as victims who needed outside help, but looking at the strengths they have to recover by themselves. This was all very pertinent to Bangladesh, and I taught short courses on this to people just back from the relief front line, dirty and tired, asking them to check out what their "victims" could do for themselves. It received a mixed response because people who suffered from a natural disaster were generally thought of as passive victims who needed to be helped by outsiders.

When I first went to Bangladesh, an American "old hand" told me that Bangladeshis were the Italians of Asia – volatile, emotional, argumentative, but also cultural, patriotic and artistic. I found all of this to be true, and Bangladesh gave me a tremendous backing in NGO/CSO 101,[4] which hopefully I was able to contribute to further.

What did I learn from the coalface?

In collaboration with the Bangladeshi CSOs, we all learned what are the essential elements of a civil society organisation, what standards we should expect from it, and how it should relate to foreign donors. There were, not surprisingly however, many kinds of CSOs. This period in Bangladesh coincided with the collapse of the Soviet Union, and equally the collapse of the Communist Party of Bangladesh. The Peasant Front of the Communist Party had many similar objectives to the more radical CSOs in Bangladesh, but it believed in committed membership in the organisation, taking power in local government, and it forswore any assistance from outside, particularly foreign donors. Anguished reappraisals took place as the Peasant Front thought how to re-design and re-brand itself to deal with the same poverty and marginalisation that the CSOs addressed.

Similar reappraisals took place as CSOs battled with the ideas of business and capitalism, at one time an anathema but now gradually being accommodated. Money lending, for instance, once the despised tool of the capitalist class and the reason for downwards spirals of poverty, was now accommodated in microfinance. If the tool was used well, and with integrity, it could be the basis of an upwards spiral of self-help.

Donors in Bangladesh generally thought that they knew better what was good for the country and good for the CSOs, and behind their arguments they always had the power of their funds. I attended many valuable and informative discussions between CSOs and donors, but in too few cases did the arguments of the very competent CSOs prevail. Usually, donors wanted to support the government, claiming that the CSOs were not sustainable, instead of working to help their sustainability.

What happened to it all?

Bangladesh has developed substantially. Its scores on various development league tables are very impressive, and much of this has come from the work of CSOs, which have, in different ways, managed to sustain themselves.

CSOs/NGOs have lessened their dependence on foreign funding principally by working on microfinance, where services in lending allow them to attract a small return in interest which can fund the organisation's existence. The bigger NGOs – BRAC, Grameen Bank[5] – have grown exponentially and have brought their competence and professionalism to many Asian and African countries. The battles between government and NGOs/CSOs have continued without change, unfortunately. The PRIP Trust has continued to the present under the management of Aroma Dutta.

Government, CSOs and NGOs have learned more about disaster relief management and disaster preparedness. All the recent natural disasters have been much less destructive than previous ones because of such preparedness.

OCA – to find out what capacities NGOs and CSOs need and how to get them to them – was pioneered in Bangladesh and is now a standard tool for working with CSOs (see Figure 10.1). There is now a large number of training courses from INGOs and local NGOs to build capacity where a weakness has been identified. This is now standard for INGOs that specialise in CSO support, like Pact.

And, on a personal note, my wife was now working for UNICEF and my three children spoke Bengali to each other, and occasionally English to their parents. I eschewed four-wheel vehicles and rode a motorbike through the dense, dense traffic, and lived to tell the tale.

Organisational Capacity Assessment Tool (OCAT)

How do you score your CSO against these model statements (1 is high, and 5 is low)

1. Governance

1.1. Executive Committee/Board/Trustees

1.1.1.	An independent governing body (Executive Committee/Board/Trustees) provides oversight to the NGO	1 2 3 4 5
1.1.2.	The Executive Committee/Board/Trustees make policy for the NGO	1 2 3 4 5
1.1.3.	The Executive Committee/Board/Trustees represent the interests of the constituency	1 2 3 4 5
1.1.4.	The Board helps the NGO with fundraising, public relations, lobbying	1 2 3 4 5
1.1.5.	The Board makes sure that the NGO's activiies reflet Board policy	1 2 3 4 5

1.2. Vision/Mission

1.2.1.	There is a clear and understandable vision and mission for the NGO	1 2 3 4 5
1.2.2.	The vision and mission are clearly understood by the staff, the Board/Executive Committee/Trustees, the constituents, the volunteers, and sympathetic outsiders	1 2 3 4 5
1.2.3.	The activities of the NGO reflect and focus the vision and mission of the NGO	1 2 3 4 5

1.3. Constituency

1.3.1.	The NGO has a recognised constituency	1 2 3 4 5
1.3.2.	The NGO has regular and participatory links to its constituency	1 2 3 4 5
1.3.3.	The NGO have the constituency to manage their own affairs	1 2 3 4 5
1.3.4.	The NGO recognises its constituency as partners in its work	1 2 3 4 5
1.3.5.	The NGO combines advocacy for its constituents with its service delivery work	1 2 3 4 5

1.4. Leadership

1.4.1.	The NGO is clear about the functions of the Director and the functions of the Executive Committee/Board/Trustees	1 2 3 4 5
1.4.2.	Decisions are clearly communicated to these they affect	1 2 3 4 5
1.4.3.	Leaders take decisions after consultation with those who will be affected	1 2 3 4 5
1.4.4.	Leaders help staff undersand their contribution to the NGO's mission/purpose	1 2 3 4 5

1.5. Legal Status

1.5.1.	The NGO is legally established	1 2 3 4 5
1.5.2.	The NGO complies with all the legal requirements of its legal identity and registration	1 2 3 4 5
1.5.3.	The NGO is aware of any concessions and allowances that it has a right to (tax etc.)	1 2 3 4 5

2. Management Practices

2.1. Organisational Structure

2.1.1.	The NGO has a clear and communicated organisational structure	1 2 3 4 5
2.1.2.	The staff of the NGO have clear job descriptions	1 2 3 4 5
2.1.3.	The job descriptions are used in staff appraisal	1 2 3 4 5

2.2. Information Systems

2.2.1.	The NGO collects baseline information aboutits constituency before starting work	1 2 3 4 5
2.2.2.	The NGO has a regular system for collecting information on its programme activities	1 2 3 4 5
2.2.3.	The NGO regularly collects information on the impact of its work following the baseline information	1 2 3 4 5

Figure 10.1 Extract from Organisational Capacity Assessment Tool (OCAT)

What were we thinking at the time?

The intense thinking and discussions between the CSOs, of which PRIP was part, led me to a substantial amount of writing that tried to capture this thinking, particularly on OCA:

Holloway, Richard. 1997. *Exit Strategies: Transitioning from international to local NGO leadership*. Pact Publications, Washington DC, USA.
Holloway, Richard. 1993. *Civil Society: The non-profit sector: trying to categorize it in Bangladesh*. PRIP and UNICEF, Dhaka, Bangladesh.
Holloway, Richard (ed.) 1998. *Supporting Citizens Initiatives: Bangladesh NGOs and society*. IT Publications and Pact Publications, Washington DC, USA. (This contains a chapter on the adventures of the Peasant Front of the Communist Party of Bangladesh.)

Others were also writing on these topics, and we were able to use their insights:

Adirondack, Sandy. 1989. *Just about Managing: A guide to effective management for voluntary organizations and community groups*. LVSC, London, UK.

Then there was the seminal work on moving from disasters to development:

Anderson, Mary and Woodrow, Peter. 1988. *Rising from the Ashes: Development strategies in times of disaster*. Harvard Institute for International Development (HIID), Harvard University, Cambridge MA, USA.

These ideas were taken up by many INGOs and UN bodies.

Ian Smillie's tremendous defence of BRAC should be read by anyone interested in Bangladesh or the complexities of working between NGOs and government:

Smillie, Ian. 2009. *Freedom from Want: The remarkable success story of BRAC, the global grassroots organisation that's winning the war against poverty*. Kumarian Press, Sterling VA, USA.

Coming to the present, there are two books which were inspired by the experiences of CSOs in Bangladesh:

Holloway, Richard. 2015. *Managing Developmental Civil Society Organisations*. Practical Action Publishing, Rugby, UK.
Holloway, Richard. 2011. The health of civil society: is this what is being assessed? In *The Future of Participatory Civil Society Assessments*. UNDP, New York, USA.

Notes

1 Bangladeshi NGOs were comprehensively known by their acronyms. The full names were rarely used, and, indeed, in some cases had been forgotten.

2 Henry Kissinger, at the time of the War of Independence, called Bangladesh "a basket case".

3 While I think the preliminary work on OCA was done in Bangladesh, a large number of INGOs and national NGOs worked out their own versions. My perspective was that they were justified if the components of organisational capacity were produced in partnership with the user organisations, but not justified if they were dictated by donors.

4 For my non-American readers, the 101 course at US universities is the basic foundational course.

5 It is important to say that the Grameen Bank fought against being called or thought of as an NGO. It said it was a bank – a very special kind of bank, but a bank nonetheless. Many of its ways of working were similar to those of CSOs, however.

11 Zambia, 1995–99

Moving into advocacy from service delivery

... In which Zambians (and Zambian CSOs) developed a second string to their bow – advocating for changes in laws, policies, practices and behaviour, to supplement their hands-on humanitarian work, particularly with PLWAs (people living with AIDS) and AIDS orphans.

Having successfully indigenised a project funded by a donor into a local trust with a local governance structure of great quality, and having trained up my deputy, Aroma Dutta, to take over my work in Bangladesh, I was ready to move on. My wife, who worked for UNICEF in Bangladesh as an Information Officer, was offered a post in UNICEF Zambia, again as an Information Officer, and we agreed to go there, this time with her as the job holder and the one with the work permit. I asked Pact if I could be useful to them in Zambia, where they had not worked previously. They said that they were ready to appoint me a Pact representative in Zambia with minimal resources for an office and without pay: I would have to earn my living by selling my services as a consultant, and separately seek to win a project for Pact from possible donors.

In the end, what happened was that I operated in Pact's name providing training courses, research and evaluation for a variety of donors in Zambia, Kenya, Uganda and Indonesia, and then succeeded in getting Pact a contract with USAID Zambia for a project on good governance. In this I was assisted by Brenda TambaTamba Liswaniso, who went on, after I left, to pioneer social accountability programmes for the mining companies in Zambia.

Good governance was a coming topic in aid circles, but no one had yet considered Zambia under this lens. There was considerable interest in its politics, but this was not yet expressed in terms of good governance. The other aspect of my work, which I found very useful, was an open invitation to devise and design capacity-building tools for Pact as an organisation to use all over the world. It gave me an opportunity to take so much of the innovative work that I had learned in Bangladesh and make it into more considered and deliberate learning materials for Pact in other parts of the world.

For me this was a return to Africa, but it was the first time working there for my wife. This was a difficult time to work in Africa: HIV/AIDS was expanding like an epidemic and ARVs[1] had not yet become common. The cemeteries on the outskirts of Lusaka expanded every weekend, and everyone you knew had lost someone to AIDS – although not all were prepared to admit it. You saw many sad and solitary people – and the implication was that they or a family member had just heard they were infected.

The World Bank was only willing to loan money to governments if they signed up to its Fee for Services policy – which meant that health services had to charge, and many people could not afford to pay. This was the case in Zambia. The Ministry of Finance had very little money from taxes and the country was desperate for aid. In this tightened situation the practices of corruption expanded greatly.

From the civil society point of view, governments were unsympathetic to the work of CSOs, considering them rivals, and the darlings of donors in respect of available funding. Governments passed legislation that, in the opinion of the CSOs, was detrimental to the poor and powerless, and CSOs found themselves faced with the need to try to get laws rescinded or replaced. This work took them into the field of advocacy, which was new for them, since most of them had previously thought of themselves as organisations that delivered services to the poor and marginalised. They started to realise that organising the poor and marginalised for advocacy was just as much of a benefit to them as delivering more material services. And this became the majority of my work in Zambia. The workshops were electric as people moved from their pride in humanitarian service provision – particularly home care of PLWAs – to realising that there was another way of working effectively, by changing the rules and making better rules which would in turn reduce the need for service provision.

This threw many local CSOs back into thinking about who they were and what they should be doing. Were they people who got together because they shared a common problem, and who wanted to alleviate this problem by collaborating with other people who shared the same problem? If so, this could give rise to mutual benefit organisations where citizens benefited themselves from the effort that they put into trying to reduce common problems. These could be drinking water programmes, but also programmes to reduce domestic violence against women or programmes to try to help youth with employment. What did such organisations require in terms of formal office-bearers, bank accounts, plans, proposals and external funding?

Or were they people who had started an association because they believed that change needed to take place, but that this change was to improve the lives of others, not necessarily their own? If this was their situation, who decided who was in need of help, and who agreed to be helped, and by whom? Everyone could agree that the disabled should be helped and would be sympathetic to an association that intended to do so, but would the same be true of drug addicts, landless people, refugees or convicted prisoners? What did such organisations require in terms of office-bearers, organisers and fundraisers, as well as plans, budgets and reports? Who would decide the activities of the organisation – i.e. how were such people chosen by others to be assisted?

Most CSOs turned out to be the latter kind of organisation (helping others, not themselves), and most of them did not have their own funds and so sought aid funding. Some even were created so that they could access aid funds.[2] For organisations that started out with the simple desire to help themselves or to help other people in difficulties, the whole issue of running a CSO became quite complex. Many people had never asked themselves on what basis they had got involved with CSO welfare and development work; they assumed they were doing good work and never considered whether they had the right to decide who got supported and how. The disabled were the people who stopped "do-gooders" in their tracks – they started to say that the only work they would agree to for helping the disabled was work that they were involved in. They would not be the passive recipients of others' charity but wanted to run their own organisations and make their own decisions.

Added to the complexities of who was initiating what, with whose support and with responsibility to whom was the question of government relations. While many CSOs said that they had good relations with district-level government officials (who often looked on them as the only source of funds that they could access), they had difficult relations with senior government officials, and frequently more junior officials were forbidden to work with CSOs unless they had specifically been given permission.

And a final level of complexity came from the relations between CSOs and the public. The CSOs which were most visible thanks to newspapers and other media were the "watchdog" CSOs that kept an eye on unacceptable government policies and practices. Such CSOs gave all CSOs the reputation of being involved in politics, and people did not easily make the distinction between politics with a small "p" – i.e. working on issues of power and decision making – and big "P" Politics – i.e. working through political parties to gain political power. Many very kind and charitable people felt that CSOs which got involved in even small "p" politics were likely to bring problems to the CSO world, were unlikely to make much impact on poverty, and, possibly, came from ideologically very different backgrounds to them. There was a lack of enthusiasm – even dislike – on the part of those CSOs working with PLWAs and with very poor people who did not have access to schools, water or food towards those which were pushing government for changes in policy and practice.

However, when care and welfare CSOs working in service delivery start to experience obstructive laws, policies, practices and behaviour that stop them from doing the effective service delivery they want to do, then they become interested in advocacy to supplement their service delivery. After considerable discussion we agreed that, for us in Zambia at that time, "advocacy" meant:

- showing that existing laws, policies, attitudes or practice are hindering rather than helping development; and
- lobbying for changes in them that will allow the CSOs (and others) to carry out more and better development work.

CSOs came to see that working in the advocacy field meant legal research, negotiating with lawmakers in the corridors of power, dealing with politicians, building constituencies for a cause, and lobbying with the public through a variety of communication tools to change the situation. This was, for many, a far cry from the hands-on personal service that was what they understood by a CSO.

Here are some examples from Zambia at this time.

Laws: Even at this crisis point for AIDS orphans, the laws on child adoption in Zambia were based on the situation of white children orphaned in the Second World War. Laws to legitimise fostering, which is the traditional way of dealing with orphans in Zambia (and indeed in many countries of Africa), did not exist, and so fostering existed in the grey area between custom and legality. Many CSOs working with AIDS orphans could see the value of getting clearer laws, but this involved a different approach to the problem than they had used up to that time.

Policies: These are instructions and guidelines within the framework of the law issued by ministries or other kinds of government agencies. In the field of agriculture, the policy of the government was for the Green Revolution approach to growing crops – purchased seeds, fertilisers and insecticides – and government support to farmers was tied to such a policy. Local CSOs suggested that the availability and cost of such inputs were both so unpredictable and so high that small-scale farmers were better advised to have a low-input approach using open-pollinated varieties, organic manure, nitrogen-fixing plants and no insecticides. Local CSOs could see, however, that in order to get a change in Ministry of Agriculture policies they would have to undertake considerable research and considerable lobbying with government officials – all of which would take time away from what they were doing. And they realised that government officials started from the premise that they were the experts, and CSO practitioners were amateur and ill-informed.

Practices: This is the situation where laws and policies do not exist, or where they are being ignored because no one is paying attention to the rights of poor or marginalised people. Such an example is the practice of the Prison Service keeping prisoners in filthy, overcrowded conditions and providing them with minimal amounts of food, bedding and clothing. Some CSOs saw that this needed reform but appreciated that they would have to expose the bad practices and try to get highly placed champions to push for better conditions – was this what they should be doing?

Behaviour: This was the most complex area for advocacy because harmful behaviour was often condoned, if not supported, by custom and tradition. It was the area, however, where Zambian CSOs saw the immediate value of advocacy and where they were most interested in doing more than palliative care. Two powerful kinds of behaviour became hot topics during this time in Zambia. The first was called "widows' property grabbing", whereby on the death of a husband (often through AIDS) the husband's family was customarily allowed to grab all the widow's property from her marriage for themselves if there was no legal will document to the contrary – and this meant land, furniture and even clothes and food. Widows and the children of this marriage, therefore, were left stranded, denuded and starving through such robberies. The second one was the rape of virgins by older AIDS-affected men in the belief that having sex with a virgin would rid them of HIV/AIDS.

Both of these were powerful issues and were kept alive because there was no law or policy against them, and they were condoned by custom and tradition. Those who were determined to do something about these harmful practices realised that help and sympathy for the affected people were insufficient. Advocacy was what was needed to try to change practices and behaviour. To women's CSOs (and it was usually such CSOs) that were angry and determined to change things, Pact tried to teach some of the mechanics of advocacy. If a service delivery CSO did decide to move into advocacy work, it was taking up a different strategy with probably a higher profile and greater risks – plus possibly greater benefits.

In tough workshops we battled together what were the best strategies: we had to point out real local examples that confirmed our points – sometimes these were accepted, and sometimes accepted only after considerable discussion:

- CSOs doing advocacy may attract opposition, particularly from powerful and well-placed men who feel threatened.
- CSOs may be accused of "playing politics", when the government has decreed that CSOs are not to be involved in politics.
- CSOs may lose track of their own strengths and mission and confuse themselves with different approaches.
- CSOs may take on extra burdens in order to do advocacy for which they do not have the skills – e.g. research, communication and dealing with the complexities of local and national government and parliament.
- CSOs may become demoralised if their advocacy campaign does not succeed.
- CSO staff or officials may be abused and insulted for doing things others consider are not their business.

Brenda Liswaniso and I spent considerable time with our CSO participants working out who were the primary and secondary targets for advocacy efforts, whether our CSOs had strong public support, and, if they did not, what they could do about it, who the CSO should ally with to make its case stronger, how to design informational materials that would have impact,[3] and other aspects of practising advocacy that had not been part of their experience to date.

On one occasion during my time in Zambia, Pact was asked by the World Bank, as part of a large review of development practices in the country, to put the case for the promotion and support of Zambian CSOs as being important and useful actors for the development of Zambia. The World Bank, as an institution, had accepted that, in general, NGOs and CSOs were useful partners for governments in development, and now it was the task of the country World Bank representatives to try to push the same position in Zambia. It was hard work. Government officials' slate of accusations about CSOs included:

- They had no right to exist ("Who voted for you?").
- They were ignorant and incompetent ("The government has trained and professional technical staff – who do you have?").
- They were subversive ("They want to change the laws and embarrass the state").
- They were insignificant ("CSOs only operate in very limited project areas").
- They were "the darlings of the donors" (i.e. donors give them funds, which they do not give to government).

And there was not much room for compromise. It should have been a Zambian CSO combatting these points, not an INGO, but the World Bank was not familiar enough with Zambian CSOs to trust them.

And to end this chapter on Zambia, I want to refer to the experience of my wife, Clare Blenkinsop, who was the Communication Officer for UNICEF in Zambia at this time. She was involved in a fascinating aspect of development practice: the use of high-profile Ambassadors to promote a particular development intervention.

UNICEF was promoting an "Expanded Programme of Immunization" globally and in Zambia, and UNICEF UK suggested that they send a British pop star to promote immunisation against polio – which was done by providing a drop of the polio vaccine in a sugar cube on the tongue of infants. And the person they offered UNICEF Zambia was Ian Dury, lead singer of the punk rock group The Blockheads. The reason for this was that Ian Dury was disabled from polio himself (caught in a swimming pool in the polio epidemic in UK in the Fifties), with legs in callipers and a withered arm. He was, however, a huge personality who had no shame about his disability. Clare was responsible for using his time in Zambia, tied in with a national vaccination campaign to try to persuade more people to take the vaccine, and, on his return to UK, to publicise what UNICEF were doing.

Clare had previously, in Bangladesh, been responsible for two global Ambassadors – Imran Khan (at that time a famous cricketer and now President of Pakistan) and Audrey Hepburn (the famous film actress), and she was now responsible for Ian Dury, most famous for coming from Billericay in Essex and for his record "Sex and Drugs and Rock and Roll", which was banned by the BBC. Ian was perhaps the least likely person to be seen in development circles (although Bob Geldof had perhaps opened the gate).

Ian asked to see the most extreme example of UNICEF's campaign to provide vaccine coverage. On the national vaccination day, Clare loaded him, his minder, the Zambian health staff of UNICEF and herself into a Land Rover and drove to the most remote immunisation post, arriving there early in the morning. Jo Bexley of UNICEF UK tells the story:

> We were standing in a clearing in the bush at about six in the morning, And there was literally just a nurse standing under a tree with a table and an icebox on top of it with these polio vaccines inside. We gradually saw people emerging out through the bush like a guy with a kid in a tin bath, strapped to the back of his bicycle, and women who had walked for three hours to get to this particular point. Ian was standing there, absolutely amazed at these people who had got up before dawn to bring forward their children and before we knew it there was a whole crowd of people lining up for their vaccine, all totally quiet, you could hear a pin trop. The Zambian doctor was asked to ask people why they had come along that day. Some of them gingerly put up their hands and said "Polio". Then the doctor said, "What does polio mean?" and one of them said, "It is when your child goes lame." Then the doctor said, "Is there anyone in this party here who you think might have polio?" and I could see another little person put up their hand pointing at Ian. The doctor said, "Even the white man can be afflicted by polio and it is really important that you should bring your children forward." I turned to Ian and said, "How do you feel about that?", and he said, "The most humbling moment of my life," and tears were just running down his face. He felt that UNICEF had made its case and he was very ready to go on radio and TV back in the UK to tell the story.[4]

Ian was getting hooked on development and prevention activities by this time and became very interested in the work that UNICEF was doing with HIV/AIDS. One of their interventions was to support a women's singing group – which they were too polite to say was formed from reformed commercial sex workers – who sang powerful songs to persuade people to adopt safe sex practices. Zambians are powerful and sweet singers. Clare introduced Ian to them and he was quickly absorbed into the group, singing along. "Oh, I'll stay here – I never want to go back to England now," he said happily. Ian also went to see the workshops making prosthetics and callipers and compared them to his. He was a unique and attractive personality.

What did I learn from the coalface?

Any society has in it people who are sympathetic and kindly disposed to others who are having a less pleasant time than they are – it may be the disabled, it may be orphaned children, or it may be those in debt, those with poor harvests, those who are sick, those who are being abused. The primary response of people to help others is very important, but with a little guided reflection, such people can see that their help can be directed towards reforming and combatting the causes and the reasons why such people are having an unpleasant time. It is important and possible to guide people in such tracks, but this is often against their personal judgement, as they are not enthusiastic to get into what they see as "politics" – because "politics" has become a dirty word in so many countries of the world.

I also learned that behaviour blanketed with the terms "customary" and "traditional" can be very vicious – e.g. the stealing of widows' property and the raping of virgins by men with AIDS, which will likely give their victims AIDS as well. It is not only very vicious but very difficult to combat by expatriates who are vulnerable to accusations of interfering in something they know nothing about. My approach was nearly always to partner with the women's organisations who were the most vocal opponents.

The return to Africa with this five-year stretch in Zambia came after five years apiece in Indonesia, Malaysia and Bangladesh. I have been asked for my views on this. At this time I thought that in Asia people's poverty was caused by, for the most part, man's (or woman's) inhumanity to other people. The effects of debt, the violence of goons and thugs, the polarisation between rich and poor were the sort of arguments I used. In Africa, however, the problems were caused more by nature – droughts, epidemics, floods – and less by what people did to each other. This, in turn, would suggest strategies to combat poverty and marginalisation. However, appalling tribal fighting in the Democratic Republic of Congo (DRC), Sudan and post-election Kenya, together with the behaviour associated with HIV/AIDS discussed in this chapter, on the one hand, and huge floods, droughts and epidemics in Asian countries, on the other, made me revise my over-simplistic thinking.

What happened to it all?

The World Bank's Fee for Service policy has been greatly trimmed back, partly as a result of the situation in countries like Zambia where poor people could not pay for services if there was a cost.

There has been a huge growth in the number of people in Zambia who have access to ARVs – but where this is something that has to be paid for, there continue to be many who cannot afford them. Where ARVs are widely available, the transmission of HIV/AIDS has been considerably reduced.

The strong dislike that government officials have for CSOs has moderated a bit but has by no means finished. For the most part, donors have not seen it as their job to defend why they are supporting CSOs while at the same time funding CSOs. In my opinion this is their responsibility, since their funding to CSOs creates ill feeling on the part of the government towards CSOs

The number of deaths from AIDS in Zambia has diminished and the cemeteries are not growing as fast.

Ian Dury died of liver cancer but left a fantastic canon of musical work.

On a personal note

USAID and UNICEF, which were funding Clare and me, were also ready to fund our children to go to international schools, which was a great relief.

Just less than 20% of UNICEF's staff in Zambia had died, likely of AIDS, during Clare's five years there.

Brenda Liswaniso, my Pact deputy, went on to be employed by a copper-mining company in Zambia to introduce corporate social accountability practices.

This was a time of expansion of internet sites. Zambian organisations (I am thinking of one in particular, Children in Need) were astonished to receive funds from overseas without asking for them – on the basis of their descriptive websites about their work – such were the feelings being generated about HIV/AIDS.

What were we thinking at the time?

As with the last chapter, I found myself writing manuals and sourcebooks on advocacy inspired by the work with CSOs in Zambia. Some of these were written while I was in Zambia, and some saw the light of day subsequently.

Holloway, Richard (ed.). 1996. *Learning from Each Other: Report of an NGO consultation on community-based approaches to development.* UNICEF, Lusaka, Zambia.

Holloway, Richard. 1998. *Establishing and Running an Advocacy NGO.* Pact Publications, Washington DC, USA.

Holloway, Richard. 2002. *When Your Work Is Blocked: Adding advocacy to service delivery.* Pact Publications, Washington DC, USA.

Holloway, Richard. 2004. *NGO Corruption Fighters' Resource Book: How NGOs can use monitoring and advocacy to fight corruption.* Impact Alliance, Washington DC, USA.

And if anyone wants to know more about Ian Dury:

Balls, Richard. 2000. *Sex and Drugs and Rock 'n' Roll: The life of Ian Dury*. Omnibus Press, London, UK.

There's also a 2010 film about his early life called *Sex & Drugs & Rock & Roll.*

Notes

1 Antiretrovirals (ARVs) are a medicine that can stabilise AIDS, but cannot cure it. At this time it was expensive and not always available, particularly to poor people and in poor countries.

2 One infamous organisation in Bangladesh was called "The Association for Blind Children" but actually traded second-hand cars, using the NGO title to get tax relief.

3 There was a publicly displayed poster at the international airport in Lusaka intended to stop Zambian travellers overseas from contracting HIV/AIDS and bringing it back to their families. It showed the family waving goodbye to the businessman at the airport with the slogan "Think of them when overseas". Once this was displayed and discussed, it was obvious how ineffective it was, but designing public posters about stopping raping virgins in the belief that it cured AIDS was a difficult task, particularly in a country where the President could only bring himself to refer to HIV/AIDS as "that disease" and where people were even very hesitant about admitting to its existence.

4 See www.developments.org.uk/articles/sex-and-drugs-and-polio/.

12 CSOs everywhere, 1990 to the present

Trying fundraising and resource mobilisation, not donor dependence

... In which CSOs learned to appreciate the limitations of foreign funding and were shown some of the alternatives to this for financial self-reliance and sustainability.

By 1990 in Bangladesh, I had started to realise the huge potential problem that CSOs experienced, the core of which was their funding. It seemed to be assumed all over the South that the work supported by civil society organisations would be paid for by foreigners. Those working with CSOs had become conditioned (almost addicted) to the idea that citizens who wanted to reduce poverty, and wanted to do this by working outside government and outside for-profit businesses, would have to identify foreign donors; learn their language, systems and procedures; sell their proposals; and report on their progress in ways defined by such donors.

What a huge problem financing was for CSOs and NGOs. There were plenty of bright, committed people who had good ideas about what to do to improve sanitation, drinking water, land reform, microfinance, etc., etc., but they lived in a world where they assumed that foreigners would pay for them to practise their activism. And when they did not get the funding they requested from aid organisations, they were bitter and perplexed – had not aid organisations been set up to fund people like them? Many considered that they had not learned the right language, had not got hold of the right code to unlock foreign pockets.

In many respects, the international aid business, as far as CSOs were concerned, was similar to the cargo cults of the South Pacific. In these, Pacific Islanders saw that all sorts of unrequested goodies came to their settlements in aeroplanes if there was an airfield. They therefore understood that the creation of an airfield would be enough to attract planes with cargo for them. Villages worked together to build airstrips in remote settlements, but, unfortunately, cargo did not come. In the same way, people in poor countries needed to form CSOs and then money would come to them. It was often as basic as that.

Once nations, or large philanthropic bodies, decided that it was strategically important for them to help local organisations, they set out to find such organisations and press funding into their hands. If such associations did not exist, then the philanthropists in the West taught them how to start such associations in the pattern of their own countries. I was asked by USAID to go to Bulgaria after the Iron Curtain had fallen in order to find and support CSOs. There were three kinds of Bulgarian CSOs in the Nineties: the first were ex-Communist State-owned bodies (professional associations of artists, architects, pharmacists, etc. that had been used to state subsidies). They now found themselves without income and had quickly renamed themselves CSOs. The second were CSOs modelled on US CSOs and often initiated by their US equivalents – good people using available Western money to carry out their initiators' ideas, but completely dependent on Western funding with few roots in their own society. And finally there was one lonely example of an independent, self-funded institution with its roots in the society. This was the *chitaliste*, a form of village cultural folk song and dance club with interests in local welfare, which had started historically at a time of patriotic fervour and still continued to the present. *Chitaliste* were independent, were rooted in their own society and had their own assets – very different from the cloned contractor for foreign funding that was the unfortunate pattern of NGOs.[1]

This is not to be confused with intergovernmental aid. The situation of the men and women who came to negotiate aid plans with Third World governments was much clearer – countries wanted access to raw materials in exchange for aid (see especially the Chinese Belt and Road Initiative), or wanted to assuage colonial guilt, or were the instruments of capitalist companies' expansion plans. What I am talking about here are the non-government associations set up by locals in the South – often by educated people who were dismayed and disillusioned by the development that their governments were practising with foreign funding. CSOs were meant, they felt, to be independent, non-governmental and developmental in design, and to have some roots in the country.

CIVICUS put out a book called *Sustaining Civil Society: Strategies for resource mobilization* in 1997, and this was greatly discussed in the CIVICUS assembly of that year. I offered to turn it into usable training materials for CSOs. I had always felt the lack of a conceptual framework for CSO resource mobilisation, and I started developing training courses for resource mobilisation in many places, often in alliance with the

International Fund Raising Group (now renamed the Resource Alliance). These were not so much training courses as workshops – we were all learning together and few were experts. Luckily, the river was moving my way and many were interested in exploring this field – and so during this time I was a freelance consultant on this subject based in Zambia for Pact. I travelled to meetings of CSOs in South Asia, in Central and Eastern Europe, in Jamaica and in East Africa. These workshops took the form of challenging people about who they thought they were, who owed them what, and what examples we had of people and organisations that were different from the contractors funded by foreign aid to do what foreign aid wanted done. This resulted in my book *Towards Financial Self-reliance: A handbook on resource mobilization for civil society organizations in the South*, which posed many challenging questions that few CSOs had adequately asked themselves. The book was published by the Aga Khan Foundation, CIVICUS and Earthscan, and these bodies made sure their partners read it.

The workshops were imbued with what have been called "Aha!" moments – moments when the participants realised some insights that they had not been aware of before. My credentials were good because of my experience, and I could legitimately pressure participants to honestly face up to dilemmas or contradictions that they partially felt but had not internalised. For instance, if foreign funding is good, why not rely just on foreign funding? Why worry about any other kind of funding?

There was squirming, there was denial, there was sometimes anger, as we squeezed out answers which all could (often grudgingly) accept.

- Foreign funding does not build local support for your work. Locals who assume that you have funds from overseas will never accept that you need their help.
- Foreign funding makes you politically vulnerable. Foreign funding provides ammunition to detractors, especially those in government, that you are being used politically by foreigners.
- Foreign funding contradicts a basic tenet of development CSOs – that of self-reliance. If you are not self-reliant, you cannot expect others to be.

Another "Aha!" moment came when I asked people to reflect on how Western charities and development organisations had been set up originally, and where the funds for existing international NGOs came from. We compared the ways in which funds were mobilised for libraries by Carnegie, or for nutrition and family planning by the Gates Foundation, the ways in which the cooperative movement got started and the Ford Foundation came into existence – all examples of resource mobilisation without foreign funders – and then compared what was happening in their countries. People appreciated how much came from country governments but also came to realise how much came from personal donations, in a variety of ways. It also illustrated how much charitable giving in the South came from religious impulses, but how little of this was used for developmental purposes. This is a tricky subject in an Islamic country, where charity was and is an essential part of religious duties (as "*sadequa*" or "*zakat*") and is often used just to feed beggars. A maverick Islamic group in Indonesia suggested that the collected funds should be used for microfinance – and faced many difficulties in putting this idea into practice.

Another set of ethical dilemmas came when we considered making money for CSOs from running businesses – something that had often been used as a stick to beat CSOs with by governments. In many cases, governments forbad CSOs from running businesses and making profits, because, they said, they were non-profit organisations. It took a long time to boil it down to the fact that there was no problem in CSOs making profits as long as the profits did not go into individual pockets but were used for the purposes of the CSO. I well remember a Zambian who ran a free school in a remote part of the country but could not afford to keep doing so. I advised him to ask all the parents to provide a goat along with their potential student, and for the headmaster to run a goat-raising and trading operation to make money for the school. As a peripatetic trainer I never knew whether ideas I had suggested had actually worked, but in this case I met the man later and he told me that the idea was working well, although he had to overcome suspicion by the parents that he was making money personally from the scheme. He had overcome this by making his accounts very transparent, but he said some parents asked him why he did not try to get foreign funds, "which would be easier".

A further set of ethical dilemmas came when I introduced the subject of getting funding from businesses, which was often approached from the business side as "corporate social responsibility". Many businesses, accused by the populace of being exploitative because they were primarily interested in making profits, wanted to prove that they were good citizens who were prepared to invest in CSOs in order to show it. Great was the suspicion from the side of CSOs at the idea of accepting funding from businesses; great was the energy from businesses to show themselves to be public spirited and imbued with a desire to support the poor. It was, at times, heady and argumentative stuff, because many businesses were showing their social responsibility by making tokenistic donations to CSOs and expecting the CSOs to act as advertisements for their do-goodery. At the same time, many CSOs could not believe in any good coming out of business – all social overtures were cunning ways of fooling people about their businesses' social responsibility and making sure that they could continue to make profits.

"Profits": it was astonishing how misunderstood this was, both from the perspective of CSOs and from that of government. In my opinion, businesses understood profits but were ambivalent about how they interfaced with social responsibility. Trying to separate income from profit was an emotional battle. It seems simple to clarify that profits were what remained when running costs were subtracted from income, but, as many CSOs clearly argued, it all depended on what you tried to cover in "running costs". If you considered that salaries were part of running costs, then the conspicuous and expensive lifestyles of many businesspeople seemed to CSOs to be evidence of capitalist exploitation, particularly if you added in that many businesses used their income to bribe government officials in order to expand and increase their profits. CSOs, not surprisingly, felt that taking money from businesses was taking tainted money. This was seen at its most extreme in Jamaica, with the Rasta-induced concept of "Babylon". All money acquired through capitalism was execrable and had to be rejected. I had real difficulty persuading people in many countries that the Ford Foundation was entirely separate from the Ford Motor Company. Its initial capital had, it is true, come from the Motor Company, but it was now legally an entirely separate body, uninfluenced by considerations of selling cars.

A step further was seen in the Philippines, where the Communist rebels had decried for a long time the ruthless and exploitative nature of Filipino businesses. This resulted in what I regarded as a fine institution, Philippine Business for Social Progress (PBSP), which was a foundation created by about 150 Filipino businesses that all gave 1% of pre-tax profits to an independent developmental organisation that ran a large number of very valuable interventions in the Philippines, both at times of disaster and at times of usual poverty. The reason, however, why the Filipino businesses had started PBSP was their fear of civil war. In their opinion, this was likely as the Communist People's Army accused the hugely wealthy corporations in the country of exploiting the people for their own profits. The Filipino businesses wanted to show that they were as public spirited as the Communists, and so they formed PBSP – and loudly proclaimed what they were doing to make sure that the people of the Philippines knew about it.

It was good to compare this with BRAC in Bangladesh, which was an NGO and a business at the same time. It started and ran a variety of businesses that made money from the resources of Bangladesh (e.g. potatoes, silk, handicrafts) and services that Bangladesh needed (e.g. educational textbooks, microfinance). The income from these businesses over 20 years enabled BRAC to become financially self-reliant. The big difference between BRAC and other CSOs was that BRAC was run by a person with experience in business, whereas so many other CSOs were very ignorant (and suspicious) of how businesses were run.

Another clash of opinions in these workshops arose over the value and ethics of taking money from government. It was rare for any CSOs to be funded by their governments (CAPART in India was a notable exception[2]), but there was occasionally an opportunity when the objectives of government and CSOs were aligned (research, for instance, particularly in remote places where government was not active, or surveys of poor people that government tended to ignore). In theory, CSOs could be contracted by government to do their work, but many CSOs felt very unhappy about doing something that would present the government in a good light, apart from the difficulties of keeping themselves separate from the usual government corruption when doing such work.

The part of the workshops that most fascinated me was when we talked about raising money from the people of the country, as opposed to receiving money from the people of other countries, either directly through donations or indirectly from government transfers that originated in the taxes levied on them. "Why on earth should people in your country give funds to our organisations?" was a striking query – and often it was a question that had never been asked because so many CSOs in the South simply assumed that foreigners (individuals, organisations, governments) would be the source of future income. An important new kind of thinking came out of these workshops: that soliciting funds from local people for your organisation was not asking them directly for money but offering them an opportunity to be involved in something that interested them, but which they could not spend their time on. If you could explain the nature of the problem in ways that resonated with people, and if you could explain your organisation as one that was competent to do something about it, you stood some chance of your own people supporting your work.

This approach involved a paradigm shift for many people. Their organisations had English names, they wrote project proposals in

English for foreign organisations, and they emphasised aspects of their work that they hoped would appeal to foreigners. We were suggesting to them that they should have names that attracted their own people, that they should explain to their own people that they were doing valuable work, and that they should suggest ways in which their own people could help them. The practical end result of these discussions in the workshops was for them to write a fundraising brochure for their own people. We suggested that it was in their own language and contained:

- their name
- a logo or photo
- a mission statement
- a slogan
- a description of the organisation
- reasons why they considered such an organisation to be important in the country
- a case study showing a successful intervention by the organisation
- endorsements by people who would impress the citizens
- suggestions about how people can help
- addresses and bank account details.

It was like getting water out of a stone, in many cases, but was often the part of the workshop that was evaluated the highest. Many were the arguments as to why such a brochure should be written in English; many were the reasons why a slogan was not necessary; many were the problems of finding someone reputable in your own country to endorse your work. In the debriefing at the end of the workshop, a substantial number of participants said, "Thank you for the insights, but we will continue looking for foreign funding – there is more chance to get such funding than to get it from our own people." A difficult pill to swallow was the issue of the organisation's credibility in its own country. We pointed out that if a scandal affected your organisation, if it was too closely linked to a political party, if it was seen as a vehicle for a particular person's glory, people would not support you. A particular facer was the issue often called "the Pajero and PC syndrome". If your organisation is conspicuous due to its expensive four-wheel-drive vehicle (Pajero) or the number of computers (PCs) it has, the public may not want to respond to its appeals.

A marvellous example of this occurred in Indonesia in parallel to a workshop. The head of a local anti-corruption CSO (Indonesian

Corruption Watch) was being interviewed on TV by abrasive anti-CSO government officials who wanted to show him as corruptly rich. The Director said, "I have a six-year-old car, a thousand dollars in three savings accounts, four cows, ten goats and 20 chickens, and one acre of land." He did not need to challenge the government officials since the audience of the TV show knew that they would either lie or be condemned from their own mouths by the amount they owned.

Once people in the CSO community had accepted the value of persuading people in their own country to support their work, there were many discussions about how to do such work – fundraising in person, at the work place, at events, through direct mail, through the telephone, through legacies, through special events (e.g. UNICEF persuaded British Airways to ask their customers to donate loose change from the country they had just left) and through the media. There were a few organisations set up to teach CSOs how to do these things (e.g. the Resource Alliance and the Red Cross), but many many more, unfortunately, who saw their job as being to teach CSOs how to write proposals to foreign donors.

What did I learn from the coalface?

- How little CSOs had even thought about the question of sustainable funding for their organisations, and how prepared they were to keep on assuming that foreigners would keep funding them ad infinitum.
- How few were prepared to confront the fact that their organisations were largely unknown in their own countries, or, if not unknown, at least little respected.
- How little businesses respected the work of CSOs. They were often persuaded to fund CSOs for political reasons or reasons of self-interest, but businesses, which above all things appreciated being business-like, saw most CSOs as well-intentioned but bumbling amateurs who were unlikely to make a big impact on poverty in their country.
- How many possibilities there were for CSOs to get involved in earning income if they were serious about starting a business that generated income – either linked to their mission or possibly completely different from their mission. To do this, however, they had to have an objective view of income and profit and be prepared to get advice from businesspeople. A very large gap in the services that donors lent to CSOs, along with their money, was education about earning money for the organisation and running a business.

What happened to it all?

More and more CSOs have reconciled themselves to living with capitalism and are prepared to take the funds that businesses offer them, or try to run an income-generation enterprise themselves. Sometimes this has come about perforce when the government of a country stops a local CSO from receiving foreign funding. Proshika, a large Bangladeshi CSO, was stopped from receiving foreign funding because the Director was accused by the government of the time of getting involved in grassroots politics. Instead of being funded by foreign donors, it was soon operating at a reduced level on the basis of a new IT business; microcredit, which allowed income from the repayment of interest; and rent of some of its property that it had acquired during its rich days (see Figure 12.1).

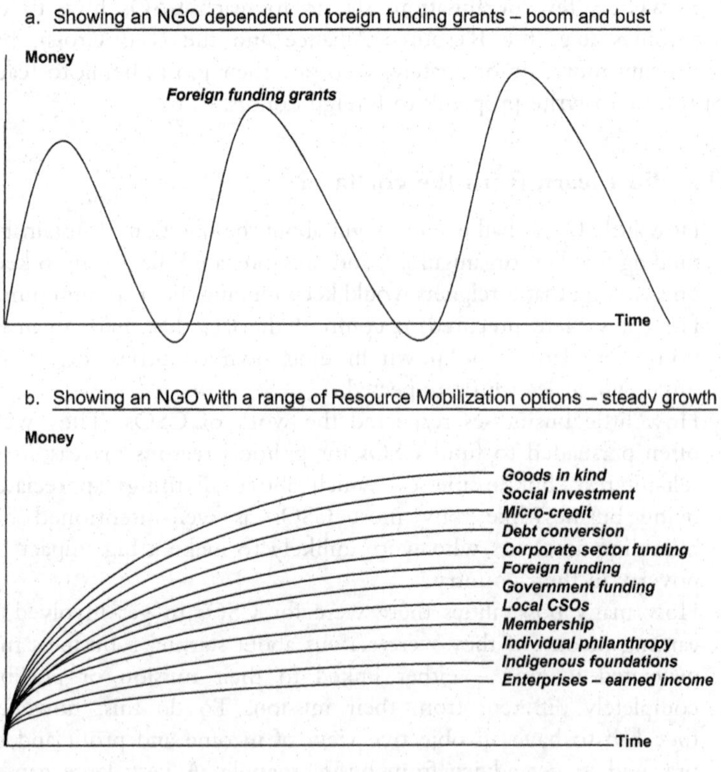

Figure 12.1 Single and multiple approaches to resource mobilisation

There has been a blurring of the edges between business and CSO in two forms: social enterprises, whereby a business has both public and private objectives and tries to make money for its public philanthropy from its private enterprise; and big businesses setting up trusts and foundations from their profits which will have a public purpose.

There has also been a lot of interesting experimentation in leverage – whereby certain profit-making enterprises are designed so that they have a powerful impact on a social ill or problem, without taking money from the enterprise to give to a CSO – good examples include the companies making fuel-efficient stoves or solar cookers, both of which enable women to cook with small expenditure on fuel, or with little harm to their eyes and respiration.

There have been some attempts to collect the huge amounts of money that have been confiscated or blocked in foreign bank accounts and make them available for quality social purposes. This involves sequestered funds from corruption (e.g. Kenyan money in Swiss bank accounts) or a pragmatic recognition that governments cannot pay back foreign currency loans they have taken from other governments, and so such loans can be written off by being paid at a reduced value in local currency to a local CSO (e.g. the Government of Seychelles and the World Bank to a local environmental NGO).

There has been considerable donor fatigue, sometimes linked to an increase in cynicism; this has given birth to a number of organisations that encourage CSO transparency by publishing information on CSO performance and funding, and on what they actually spend their money on.

And from a personal perspective, looking at the three sources of support to CSOs brought me to this thinking: those who enter the track of the commercial sector are rarely sensitive to the problems of the poor: those who enter the track of the non-profit, non-government sector are rarely sensitive to the issues of a capitalistic system of wealth creation: and those who enter the track of the government sector are rarely well informed about the damaging nature of corruption on a country – instead seeing it as a lubricator of development. There is a huge need for schools and universities to provide a Development Studies curriculum for secondary school and university-level students to try to get these three tracks to understand each other.

Clare and I had two children by this time, and, thank the Lord, home help to help with looking after them – since I was frequently away on training courses which lasted seven to ten days.

I said at the start of this book that I prided myself on being down at the grassroots; doing staff training courses was a bit different, but once we had worked out the kinds of questions that we needed to ask in order to understand how they understood themselves as CSOs, and how they were regarded by government, the public and the donors, a tremendous amount of material was harvested about sustainability and how to keep going when the donor funding dried up. It involved some pretty grassroots discussions.

What were we thinking at the time?

The workshops generated a lot of thinking that I wrote up at the time, principally in:

Holloway, Richard. 2001. *Towards Financial Self-reliance: A handbook on resource mobilization for civil society organizations in the South. Aga Khan Foundation.* CIVICUS and Earthscan, London, UK and Sterling VA, USA.

Sources of information on resource mobilisation came from the Resource Alliance (www.resource-alliance.org), which started life as the International Fund Raising Group, and the magazine, *Alliance* (www.alliancemagazine.org), bringing information on who was doing what in philanthropy. Both of these have continued and prospered, and have expanded their reach.

The Center for Civil Society Studies at Johns Hopkins University has been involved in Resource Mobilisation work from this time to the present and has, through its Director, Lester Salamon, regularly produced valuable information. In 2014 he brought out a book expanding our understanding of the practice still further, and, more recently, work claiming, for philanthropic purposes, illegal funds that had been secreted overseas.

Salamon, Lester. 2014. *Leverage for Good: An introduction to the new frontiers of philanthropy and social investment.* Oxford University Press, New York, USA.

Notes

1 *Chitaliste* were written up in the book *Beyond NGOs*, which includes many such examples. Holloway, Richard. 2007. *Beyond NGOs: Civil society organisations with development impact.* Aga Khan Development Network and INTRAC, Geneva, Switzerland.

2 CAPART, the Council for the Advancement of People's Action, is an institution under the Ministry of Rural Development of the Government of India to fund Indian CSOs working in rural development.

13 Indonesia, 1999–2004

Never again, neither Suharto nor his corruption

... In which, having joined and left a USAID project, I joined a national organisation for Reformasi *(Reform) after 32 years of President Suharto. How this organisation helped me to understand the depths of corruption in all walks of Indonesian life, and the difficulties of doing anything about it, and offered me the opportunity to document good governance.*

I was brought from Zambia after five years to Indonesia again, but this time to Jakarta. The *Orde Baru* (i.e. the old regime of President Suharto) had ended in 1998, and a new President Habibie, shortly followed by President Abdurrahman, took over. A whole raft of new laws had brought in the era of *Reformasi* (Reform), including handing back East Timor to its citizens. It was a new and exciting time to be in Indonesia and people felt that they were breathing different air. The Citizens' Parliament had stated the opposition of the Indonesian people to corruption, and the control on NGOs had been virtually lifted. I came into Indonesia to do one job, but soon resigned and moved into another one that allowed me to be part of the exhilarating *Reformasi* process.

The employer for the first job was one of the organisations crudely called a "Beltway Bandit", by which is meant one of the US for-profit contracting companies that aggressively seek USAID funding and have their office on the Beltway in Washington DC (an American equivalent of a ring road). I openly admit my bias – such organisations are profit-driven and yet work increasingly through non-profit organisations – and in my opinion there is a clash in their motivations. Their heart, in my opinion, is not in the non-profit or civil society field. It was significant that the previous contract that this organisation had had with USAID in Indonesia had been for a capital city waste disposal facility. They had never worked with civil society or on good governance issues.

The programme, from my perspective, was simply about USAID showing how much it was prepared to spend on CSOs in Indonesia – positioning themselves for *Reformasi* and the regime change in the country, and getting close to the State Department's interpretation of US interests. In my opinion, the integrity of the grants, the value of the organisations receiving grants and the impact expected were less important to the organisation than being alongside USAID. And nothing had changed about the extraordinary list of conditions for any organisations receiving USAID grants – work on "weather modification" (see Chapter 10) was still prohibited for CSOs!

I worked for one year and then resigned – a tricky decision because I had nothing else to go to.

However, luckily for me, a very new organisation was being created in Indonesia by Indonesians strongly motivated by *Reformasi*, and keen to prevent any return to the bad old days of Suharto. This new organisation was supported by the World Bank, the Asian Development Bank (ADB) and the UNDP, but it was governed by Indonesians and was a non-government entity. It was the Partnership for Governance Reform in Indonesia – whose name clarifies the context. After 30 years of increasing autocracy and corruption, there was a critical mass of people who wanted substantial changes in the way in which Indonesia was governed. In many cases these changes came about through laws from a reformed national parliament and via the decentralised provincial parliaments, but the technical advice on how to make the new laws work was offered by the Partnership, and this was a non-government, non-profit body.

An official paper in 2000 stated:

> The Partnership for Governance Reform in Indonesia emerged from a common consensus of the Indonesian Government and donors at the 2000 CGI [Consultative Group on Indonesia] meeting. At the time it had become obvious that sustainable reform of governance practices cannot be affected by "conditionalities" imposed from outside, but rather must be based upon building a domestic demand for such reform coupled with supporting the implementation of reform within the institutions concerned. National ownership is a key to change. The Partnership has come to phrase the notion of building up demand by the citizenry for reform with a commitment by institutions to implement reform as "pressure from without, capacity from within." To be successful therefore there must be a strong Indonesianisation of these

> concepts and processes. This means incorporating the ideas within the context of real experiences and traditions within Indonesia. This is most crucial to guarantee national ownership and ultimately to both the viability and the sustainability of the reforms themselves.[1]

I was fluent in the Indonesian language by this time, and it was a thrilling opportunity to be involved with strongly motivated Indonesians in something that so clearly spelled out what the Indonesians wanted to do, and what was wrong with the ways in which donors usually wanted to work.

Right from the start its support was agreed by the three large bodies of the World Bank, ADB and UNDP, and it was not long before other national aid organisations from other countries (UK, the Netherlands, Sweden, Australia, the EU, etc.) wanted to join as well. International donors had become very distrustful of supporting anything managed by the government and wanted to move into supporting something that was national and covered the whole range of governance, not just the more specific sectors of larger NGOs. Donors, however, had to understand that the Partnership set the conditions.

I was recruited by the World Bank for the position of NGO and Anti-Corruption Adviser and contracted by UNDP. The quality of the other staff was very high, with people taken from CSOs, universities and the media. The Partnership's work took place in a collegial atmosphere of commitment to a new Indonesia, and with a determination that Indonesia would not drop back again into a regime like that of Suharto. It was heady times in Indonesia: huge marches were taking place of citizens for and against possible Presidents, and for various rival candidates – and these happened just in front of our office on Jalan Thamrin in Jakarta, blocking the six lanes of traffic of a dual carriageway.

There were different CEOs of the Partnership, but during my time it was led by a remarkable woman called Felia Salim, who was a senior banker in Indonesia for many years, Director of the Jakarta Stock Exchange and on the boards of Transparency International Indonesia and the TIFA Foundation, set up by Open Society Foundations. She illustrated the quality of people who offered themselves to the Partnership, which was seen as an important organisation at a time of crisis in the country. All of them wanted to see Indonesians in the driving seat, telling the donors what needed to be done and not just responding to donor suggestions and conditionalities. As an aside, she was also the only person in Indonesia, as far as I know, who had a women-only gamelan orchestra in her house.

The Partnership was committed to the following principles:

- Speaking with an Indonesian voice.
- Involving multiple stakeholders in governance reform (government, business and civil society).
- A two-pronged approach of building consensus among people committed to governance reform and then supporting the work of people and organisations committed to action.

The business of the Partnership was carried out in Indonesian (which I had learned during my time in central Java with Oxfam in 1979–84) and translated where necessary for the donors into English. At a very early meeting with her staff, Felia Salim pointed out the striking fact that, whatever we wanted to do, our name could not be translated into Indonesian – and this illustrated the nature of the task we were facing. In Indonesia, government translates as "*pemerintah*", and this reflects the mindset of Indonesians – that government is the source of all authority. The name "A Partnership for *Governance* Reform in Indonesia" suggests that all authority does not come from government alone; there are a number of sources of decision making that can be involved in governance. But there was no popular word for "governance" in Indonesian, and, as Felia Salim told us, we had to educate people about what governance means. On a linguistic level, adaptable Indonesian with its prefixes and suffixes allowed us to produce "*Kepemerintahan*" to mean governance; this became part of our logo, and it was our job to popularise it.

Such a linguistic conundrum applies to many languages where there has been a strong centralised government. When you are looking for a word that encompasses all the decision-making bodies in a country – government, business, civil society, religion, tradition – it is not so easy to find one. In so many countries, central government has put its stamp on our thinking, our language and our mindset.

The Partnership had taken on for itself eight sectors that it considered of the highest importance. As you can see, this was not just a CSO with a thematic interest; this was an organisation interested in the whole field of governance reform in Indonesia:

- the legislature
- the justice system
- elections
- the Civil Service
- security (including the police and the military)

- corruption
- decentralisation
- business.

My work primarily dealt with corruption and there were Indonesian linguistic issues with this, which again illustrated a mindset problem. Corruption in Indonesian was *Korupsi*, and it was nearly always linked to collusion and nepotism – which made a natty acronym of KKN (*Korupsi, Kolusi, Nepotisme*). This was the standard and official way of referring to corruption. The fact that these well-known traits in Indonesian life were referred to by Indonesianised English words was bizarre, and another example of the "mindset" problem mentioned above in relation to governance. The World Bank's definition of corruption is "the abuse of public office for private gain", and this was understood in Indonesia as government officials making illegal money from their office. But there was no word for this, apart from the Indonesianised English word "*Korupsi*". There was the simple Indonesian word for stealing, "*pencurian*", but Indonesians did not easily conflate corruption with stealing. A very regular observation in Indonesia is that a person who steals a chicken in the market is attacked and punished, but a Government Minister who steals millions is immune. It is my feeling that retreating into the use of "KKN" allowed (and allows) Indonesians to consider corruption as something different from stealing, and something that should be dealt with differently, if dealt with at all.

There was a lot of interest in overcoming corruption in Indonesia, and I was very well placed to have a hand in that. The way in which it was considered by the whole range of Indonesian people was fascinating. In the words of Robert Klitgaard of the World Bank, it was "an open secret" – something that everyone knew about but few were prepared to talk about officially. The fee for a government service was publicly displayed, but everybody knew that this was not the reality of what was required. Most people knew the actual situation, and – surprisingly to me – went along with it. I was very interested in understanding what people actually did in respect of corruption and whether they appreciated its harmful effects, rather than in receiving opinions generated by scholars, political scientists and intellectuals.

With the help of the World Bank I was able to achieve the following within the work of the partnership:

- A survey covering 2,600 respondents from 14 provinces on attitudes to and experiences of corruption. This revealed the strength of feeling of most Indonesian people about corruption and their clear identification of the most corrupt organisations, but also the ambivalence that many had about what corruption is acceptable and what is not. This was carried out by a market survey organisation, more usually involved in assessing attitudes to new toothpastes, and was right down to household level.

- Six regional anti-KKN meetings of about 100 people each, drawn from the executive, legislature, judiciary, universities, NGOs, traditional organisations, religious organisations, trade unions and businesses. These took place in major regional capitals. The meetings were to pin down what citizens felt about "Reforming the Civil Service and the bureaucracy"; "Changing attitudes and behaviour"; "Reforming the justice system and essential laws"; "Promoting popular pressure and citizens' oversight of government"; "Providing education and information about how corruption works"; and "Reforms needed in key institutions". We also asked for suggestions for actors and activists who could do this work.The meetings were fascinating – on the one hand there was an opportunity for people to blow off stream about egregious incidents of corruption that they knew of in their locality, but on the other hand there was a reluctance to believe that any measures we could suggest would make any difference, even if they were very clear that this was what needed to be done. The impunity of so many people over such a long time was confronted by the enthusiasm for *Reformasi*, and their acceptance of impunity usually won. It was clear that *Reformasi* was not *Revolusi*.
- The six regional meetings produced a second national meeting, which, under strong pressure from the Partnership, produced 20 practical ideas.

We also produced two books in English and Indonesian, illustrated with cartoons, commissioned from writers, journalists and development workers. These were intended to clearly say what corruption meant at the grassroots level:

- *The Poor Speak Up: 17 stories of corruption*
  Very specific corruption cases were gathered from real-life experiences in Jakarta, Yogyakarta and Makassar with a final chapter on "How corruption affects the poor, and what to do about it".
- *Participatory Corruption Appraisal: A methodology for assessing how corruption affects the urban poor*
  This methodology allowed outsiders working with insiders (i.e. CSOs familiar with the communities that are affected by corruption) to understand and systematise information about corruption in poor communities. It also allowed insiders, with outside assistance, to

think through what they might do to improve their situation. It used focus group discussions very familiar to CSOs working with the poor, and tried to analyse the reasons for corrupt behaviour.

We also produced, with a local research organisation called the Aksara Foundation, a four-volume collection of 15 research studies of corruption in Indonesia called *Stealing from the People*. This collection of articles was intended to document the actual extent of corruption and force people to see its impact. In particular, I wanted to assemble material that was able to overcome the argument made by many that "petty corruption" is not that significant. One article looked at the amount of the illegal levy paid on each transaction at the vehicle licensing office, then multiplied this by the number of illegal transactions made every working day at one such office, and by the number of such offices throughout Indonesia. The sums were enormous. Another article looked at illegal payments to the police and the percentage of such payments that were transferred up the police hierarchy to police supervisors, inspectors and higher officials. Again, the sums were enormous. A third chapter looked at corruption in Indonesia's foreign aid: "Naivety, complacency, and complicity within donor agencies".

One illustration of the extent of corrupt behaviour was in the compilation of the book itself. A consultant who seemed to have the right credentials for a chapter on local government corruption was contracted, only for me to find, when he provided the manuscript, that it was stolen and plagiarised from another author. We stopped its inclusion but could not get back the advance on the payment to him.

All of this work with the Partnership educated me – and, I hope, many others – on the nature and extent of endemic corruption in a country. The education was not, however, restricted to the work of the Partnership. Every day, examples of further and deeper corruption were revealed by the media, in particular *Tempo* magazine:

- At the terminal in the national airport, Indonesian workers returning from overseas employment continued to have their savings extorted from them by criminal immigration and customs officers.
- Senior government officials would make sure that envelopes full of money were openly left on the desks of Members of Parliament with encouragement to them to vote for a particular bill that would give an opportunity for illegal income to come to the government department that they represented.

- Judges were open to the highest bidder for their judgments, and the bribes were managed through their clerks. This was well documented by a local CSO, Indonesian Corruption Watch.
- Budgets for government departments were officially divided into routine and development expenditure, but the funds were fungible and were not able to be understood by the public.
- Civil servant positions were bought and sold, and the prices were dependent on the amount of corrupt income that could be expected. Ministries with many opportunities for corruption were called "wet" ("*basah*" in Indonesian), an example being the Ministry of Works, and those with few opportunities for corruption were called "dry" ("*kering*"), such as the Ministry of Sport.

I came into the field of corruption studies through my work with NGOs and CSOs – they were the kind of people and organisations that knew how corruption worked because they were acquainted with the poor and marginalised and were prepared to talk about it. I was astonished at the size and pervasiveness of corruption, and particularly how it led to the polarisation between rich and poor. After Indonesia I approached every country I worked in with a different mindset, conditioned to look for examples of rich and powerful people messing up the lives of the poor and powerless, and getting away with it ad infinitum.

The Partnership gave me the opportunity to edit its first governance report, which brought together all the different elements of governance that it was interested in. It was astonishing, and depressing, how corruption underlined nearly all of these elements.

What did I learn from the coalface?

- Most people, rich and poor, in an endemically corrupt country, practise corruption. It is only when corruption is extreme, and beyond a limit that is difficult to identify, that people feel strongly enough to try to do something about it. And the rich and powerful are quite able to employ thugs and gangsters ("*preman*" in Indonesian) to stop those who want to do something about it.
- It is very possible when sitting opposite someone from government and discussing the evils of corruption that this person is the recipient of the "vertical hierarchy of corruption" – i.e. that he or she is receiving a percentage of any illegal extortion that subordinates have received.
- Amazingly, very poor people, instead of being angry at extortion by other poor people who happen to have more power than

them, can explain that such people are poor as well and need to receive bribes for them and their families to live at even a low level, and they pay up.
- Indonesians were not prepared to call out corrupt officials to their face.

What happened to it all?

Soon after I left the Partnership and went on to my next job in East Timor, the Anti-Corruption Commission in Indonesia was established with serious and solid staff, and a legal set-up that seemed to allow it to work with integrity. As its work started to bite, however, it received vicious, false and personal attacks on its senior staff from those they had accused. Its future was precarious, but a large and widespread campaign was undertaken by CSOs to expose the accusations, and the institution was saved. It continued, under further Presidents, to do its work with integrity. If not invulnerable, it is at least able to attack anyone in the country who they consider to be guilty. Indonesia is still a country of great corruption, but the mood of tolerance has modified.

The Partnership has continued to the present and has taken on a range of other topics and issues in addition to the ones it started with. Its staff is now completely Indonesian, and it has a strong reputation for professionalism and probity.

What were we thinking at the time?

The Partnership was in business to publicise harmful aspects of corruption and the workshops and research it underwrote gave us the opportunity to do so.

Holloway, Richard and Makarim, Nono (eds). 2002. *Stealing from the People: 16 studies on corruption in Indonesia.* Partnership and Aksara Foundation, Jakarta, Indonesia.

Stealing from the People has four volumes:

1. *Corruption: From top to bottom*
2. *Big Feast: Soldier, judge, banker, civil servant*
3. *Foreign Aid, Business and State Enterprise: Counting the cost*
4. *The Clampdown: In search of new paradigms*

Hardjono, Ratih and Teggemann, Stephanie (eds). 2002. *The Poor Speak Up: 17 stories of corruption.* Partnership, Jakarta, Indonesia.

Partnership. 2001. *Diagnostic Study of Corruption in Indonesia.* Partnership, Jakarta, Indonesia.

Holloway, Richard (ed.). 2002. *Controlling Corruption in Indonesia: 20 practical ideas.* Partnership, Jakarta, Indonesia.

Holloway, Richard (ed.). 2003. *Assessing the State of Governance in Indonesia in 2002: The Partnership's governance assessment.* Partnership, Jakarta, Indonesia.

Holloway, Richard and Teggemann, Stephanie. 2002. *Participatory Corruption Appraisal: A methodology for assessing how corruption affects the urban poor.* Partnership, Jakarta, Indonesia.

There were few documents on corruption that go into the details of how corrupt people work – and there have continued to be few. I tried to get beyond that in Indonesia and subsequently in Myanmar (see Chapter 18) with regard to corruption in the timber industry.

There was, however, a fine document from the World Bank along these lines that many assured me was based on Indonesia.

Della-Giacoma, Jim and Wherry, Frederick. 2001. *The Anti-corruption Handbook.* World Bank, East Asia Environment and Social Development Sector, Washington DC, USA.

Note

1 Paper presented by H. S. Dillon, Executive Director of the Partnership for Governance Reform in Indonesia, at the pre-CGI meeting in Jakarta, January 2005.

14 East Timor, 2002–04

Moving from relief and human rights to development and civil rights

... In which two years after the crippling independence of East Timor, local CSOs moved from fighting for human rights and independence to development interventions, and advocacy for addressing their communities' problems with the new local government.

Following the end of my contract with the Partnership in Jakarta, Catholic Relief Services (CRS) offered me a job to manage its new contract with USAID[1] for a programme to support CSOs and advocacy in East Timor – the occupied territory of Indonesia, now recently independent. I stayed there for two years.

The smallest amount of briefing about East Timor (called Timor-Leste after the Declaration of Independence was achieved in May 2002) reveals the complexity and complications of this small nation situated at the end of the Indonesian archipelago. It had a strong traditional structure based on kingdoms and chieftaincies but had been invaded very many times by large numbers of varied foreigners. The latest invasion was of foreigners keen to help the country develop, and I was one of them. Tetum was the traditional language of most of the country (although there were many local languages); Portuguese was the language of the colonisers, the educated and the Church; and Indonesian, after 32 years of occupation, was also very widely spoken.

To illustrate this complexity, Timor-Leste had been subject to, over time, the following invasions:

- The Portuguese arrived in the sixteenth century to trade in sandalwood, beeswax, sea turtle shells, slaves and coffee, and established a colony in 1769. For 300 years the Portuguese used the land as a place of exile for political and criminal prisoners, but did little to develop it. They left it in 1975 following their own revolution in Portugal and their desire to divest themselves of their colonies. The Portuguese brought in substantial numbers of Catholic priests who overlaid traditional beliefs with Catholic Christianity. The Catholic Church was the only agent of education and it spread the Portuguese language among educated people. It also sent East Timorese for religious education to Portugal.

- The Australians went forward through East Timor to meet the Japanese in 1942. After guerrilla wars in the mountains in which Timorese fought with the Australians, the Japanese occupied the land for three years (1942–45) until the end of the Second World War.
- The Indonesians invaded in 1975 and stayed there until 1999. This was a bitter and destructive invasion which destroyed the lives of two-thirds of the people and killed over 100,000, either by fighting or by starvation. The fighting was between the Indonesian Army and FRETILIN, a Marxist political party supported by the people of Timor, and its military wing, FALINTIL. The Indonesians brought in their own civil service, language and government structures, and, in the fighting over independence, created internal divisions in the country. The Indonesians supported local militias who were pro-integration with Indonesia and some Timorese fought for Indonesian-supported militias. The strong Indonesian ideology of Pancasila[2] forced all its people to belong to one of five recognised religions (and no traditional religions): membership of the Christian Catholic religion expanded from 20% in 1975 to 95% by 1985. Many churches were built and traditional religions went underground. After the 30 August 1999 referendum in which Timorese voted 78.5% for independence, there was huge destruction of the infrastructure of the country by pro-integration militias helped by the Indonesian army.
- The UN peacekeeping forces (INTERFET) came in 1999 to restore peace and handed over the management of the country to the UN Transitional Authority. INTERFET was mostly Australians, but they were joined by a range of other nations – Singapore, Malaysia, Fiji, New Zealand, Portugal, Brazil, Japan and Bangladesh.
- The Transitional Authority handed over the governance of the country to the State of East Timor with the Declaration of Independence, and the country became known as Timor-Leste.

This was the last "invasion": all sorts of people came through the UN, through inter-government and non-government organisations to help Timor-Leste with its development problems. There were the usual UN agencies, coordinated by UNDP; strong representation from the Philippines, linked by a common religion in Catholicism; a slow build-up of teachers and language instructors from Portugal (the new government had decided that Portuguese was the official language and the language of instruction in schools); a range of international development NGOs; and a range of foreign solidarity organisations that had been supporting East Timor politically during the Indonesian occupation and had lobbied for its independence.

Arriving in Dili in 2002, I was aware of this huge complexity – soldiers of many nations carrying what were to me alarmingly large guns around the streets, rows and rows of burned and gutted buildings, all against beautiful seas and beaches, mountains and forests, and very poor people. There were endless enthusiastic development workers, particularly those from ex-solidarity organisations still celebrating the achievement of their successful mission and, initially, very few Timorese. They were still working out where they fitted into this swirl of interests.

As someone working with an international CSO (which in turn was working with many local Catholic CSOs), I found the government unfriendly and uninterested in our work. Diaspora Timorese political leaders from Mozambique who had been members of FRELIMO there became supporters of FRETILIN in its image back in Timor. FRETILIN was a Socialist party, evolving from a Communist party, and its leaders saw no value (in fact, they saw danger) from citizens' organisations which were not members of their one political party. To my (initial) surprise, Timor-Leste was also festooned with images of Che Guevara – on T-shirts, on buildings via stencils, on posters. People identified hugely with Cuba. I found, ready-made, the slogans (in Spanish, not Portuguese), the images and the narrative of Che Guevara. Very few Timorese, however, knew anything about him apart from his picture and the slogans.

I found the following kinds of CSOs on the ground:

- Timorese versions of Indonesian CSOs: a substantial number of Timorese activists had been to universities in Indonesia, particularly in Java;[3]
- Timorese church-based organisations often situated in (and linked to) a diocese or a congregation;
- Timorese clones of international NGOs, or bodies formed by such bodies;
- residual political/activist groups that had been supporting FRETILIN as a guerrilla army; and
- traditional Timorese clans or chiefdoms.

The first three were by far the most common; in fact, it took me some time to become aware of the last two – particularly the ruling families of the traditional kingdoms and clans (of which more later).

CRS was, not surprisingly, closely linked to local Catholic groups and CSOs, which were my first contact with civil society in Timor-Leste. CRS had been one of the first outside organisations to arrive in the country as soon as access was possible, and it had been involved in relief distribution. It was also very involved in responding to the human rights abuses of the previous 30 years, which had accelerated in the last two years with the lead-up to independence. To my way of thinking, CRS, and other foreign-funded organisations, had helped inculcate a mindset among local civil society activists, through their interaction and their gifts, that the people of Timor-Leste were:

a) victims of human rights abuses that needed to be remembered and documented;
b) victims of man-made disasters, and therefore legitimately entitled to relief supplies; and
c) newly emergent members of the modern world, and therefore entitled to electricity and all the accoutrements electricity brought – particularly computers, printers and generators.

I remember one of the first CSOs I visited outside Dili in 2002. It had received from some foreign aid organisation a huge TV screen, a computer linked to it, a printer and a collection of DVDs documenting the bad events of the Indonesian occupation, including the Santa Cruz Cemetery massacre. It also had a diesel-fired generator to make all these things work. The CSO considered that its job was to use these tools to educate and sensitise the people in its locality about the atrocities of the Indonesian period, and to encourage them to support the newly elected government. When I enquired about the problems of the people that they might want to address (such as clean water, reforestation, food crops, nutrition, cash crops and marketing), I was told that these were not their priority; moreover, they had been given all this equipment and they needed to use it. They were dismayed, however, that they had no ongoing source of diesel, which they saw as their greatest need.

This seemed to me to be retrogressive and to be ignorant of the ways in which the people, on the one hand, and the donors, on the other, would expect CSOs to work in the future. With the agreement of CRS, I wrote a series of brochures (in three languages – English, Indonesian and Tetum) to try to encourage a change in the mindset.

These became the basis of CRS's training work with CSOs. The introduction said:

> It is now 2003 – the emergency is over, and the long term development process has begun [...] During the emergency and transition period it was urgent to deliver humanitarian supplies, and to carry out civic education in connection with the elections. A lot of organisations received money with very few questions asked during that time [...]
>
> Foreign funding for handouts channelled through NGOs is not available any more. NGOs which only operate as channels for foreign funding and do not think of doing anything else are likely to die out [...] the NGOs which will continue into the future will be NGOs that work in different ways, particularly NGOs that have a very clear idea of what they want to do, and which involve villagers in their planning [...]
>
> Many NGOs are staffed by people who came from an activist background. They were determined to gain the freedom of Timor-Leste and they worked with great commitment to do so. That battle is now won, and the present battle is against ourselves to make sure that the country develops in the best way possible.[4]

One of our difficulties was that there were still plenty of donor agencies that did not buy into this form of thinking and were happy to keep supplying handouts and encouraging a handout-dependent mentality.

If we succeeded in persuading local CSOs to think in these terms, there was the opportunity to train them in managing an efficient, professional and caring organisation. The booklets and the training courses we promoted covered:

- Being an NGO in Timor Lorosae – identity and values
- Mission, vision, strategy, programmes
- Structures and systems
- Skills and abilities
- Material and financial resources
- Partnerships
- Making a budget and reporting on it
- Managing a project or grant
- Managing people
- Managing a community project
- Working with the government
- Undertaking advocacy campaigns.

Each booklet contained "Questions to ask yourself (of these subjects)" and "Questions to ask yourself about your NGO" in order to encourage CSOs to think about what it meant to be a CSO in East Timor.

The CRS programme, funded by USAID, helped citizens' organisations or associations to become competent in doing what they had identified that they wanted to do, and in general this was done with no reference to government whatsoever. There were two reasons for this: first, government offices, and government officials in general, were taking time to become established, and thus there were many vacuums in the government machinery – particularly the apparatus of local government; second, there was precious little inclination for government or CSOs to have anything to do with the other. I have already mentioned that the political ideology of the government in 2002–04 was Socialist and strongly influenced by Mozambique's FRELIMO: there were plenty of government documents which talked of the value of "civil society" (produced, in my opinion, under arm twisting by UN advisers), but by 2004 (i.e. two years after independence) the government still had not passed a law giving a legal identity to non-government organisations, nor had it established a supportive and enabling legal, fiscal and institutional environment.

Both sides played unhelpful and childish games – the government called for meetings with insufficient publicity and then complained when the CSOs did not attend in sufficient numbers; CSOs called for meetings and then complained that only low-level government officials who did not have decision-making powers attended them. High-level government officials would speak only Portuguese, which NGO people did not speak, and CSOs would speak only Indonesian, which high-level government officials very probably knew but would affect not to speak. I attended one such meeting in the brand-new government buildings, and you could cut the atmosphere with a knife. With icy politeness, each side would denigrate the work of the other. There was no populist group or individual who was able or interested to shake people out of these prejudices and seek to get some collaboration between government and CSOs, with the intention of bettering the lives of the people of Timor-Leste.

Towards the end of my time in Timor-Leste, the government had started to set up local government offices and place officials in them, and we started to teach CSOs about advocacy, to lobby for a change in law or practice, or for the provision of services. We wrote more brochures about how CSOs could lobby local government, this time in four languages, adding Portuguese. In many ways government was not sure of itself or what it was meant to provide, and so there was plenty of room for CSOs to mould government in the shapes that it would like to see. For instance, one of the exercises we suggested was as follows:

> In one part of Timor-Leste, parents were complaining that the teachers were always arriving late to school, so their children were not learning. Investigation proved that the government had given the teachers housing that was a long way away from the school, or no housing at all. The parents decided that the District Education Secretary had the power over this issue and decided to approach him in a group, to lobby him to try and get the teachers' accommodation changed.[5]

By 2004, very few citizens knew what a District Education Secretary was, and they would be apprehensive about approaching government offices in a group to ask for anything. In turn, it was quite possible that a District Education Secretary was unclear what his or her powers and responsibility allowed him or her to do.

Another aspect of establishing a local government structure was that it clashed with the traditional hierarchy of powerful people in local communities. Was the government going to diminish the power of the local *liurai* or headman, or the local priest, in favour of a government official? The World Bank introduced its model for local government: CEP – Community Empowerment and Local Governance Project. This was a model that the World Bank had used in Indonesia, then Afghanistan and finally Myanmar. FRETILIN did not like this structure and wanted something that reflected the "*região*", which were the organisational structures in the resistance. CEP was rejected.

In the first part of our work we had tried to help people understand what a developmental CSO was and could be, and then set about building its capacity to undertake the development interventions it had decided on. Now, in this second part of our work, we helped people to understand advocacy: i.e. a systematic and organised effort by CSOs to create helpful laws, policies, practices or behaviour, or to change those which were not helpful. We were then able to teach them the skills of how to do this – carrying out local research, suggesting alternatives, using media and other methods of communication, networking and coalition building, negotiation and diplomacy.

Wherever possible, given the delicacy of demanding services from government, we would urge the greatest use of self-help and voluntarism.

There were many existing traditional governance structures for rules and decision making of which foreigners were ignorant but that local people knew well. For the most part, sadly, we did not bring these into discussions about governance. An example is *Tarabandu* (taboo or ritual practices). It was accepted that a *liurai* had the power to designate a piece of land to be out of bounds to ordinary people and animals. This could be land whose water sources or whose grazing needed protection, or land connected with sacred sites that needed to be reserved in some way. What was very clear was that once such land had been identified by a *liurai* – usually by a bundle of grasses and branches tied to a tree – its authority was very strong. Everyone knew what was meant and obeyed the indigenous law.

Learning about *Tarabandu* persuaded me to find out more about traditional society and I started to understand a whole subculture of CSOs based on *Lia Nain* (sacred houses). The limited anthropological research in Timor had informed a few academics about the rich and important cultural heritage of Timorese people – the clans and lineage groups, the sacred houses where the clans' sacred objects are kept (often helmets and armour plating from old Portuguese days) and those who guard them. What was not so well known was how pervasive these cultural associations are, and how powerful.

I persuaded Timorese friends to take me to see their sacred houses, which are massive, cathedral-like buildings made of huge timbers in the remote countryside, all of which were erected by the voluntary labour of clan members. They are very impressive. I was complimented on my interest in them – there was no feeling that I was transgressing into something forbidden from which foreigners should be excluded. I found that every Timorese belongs to a clan and is very aware of this belonging and the duties involved. They are aware of their chiefs and their office-bearers and their responsibilities within the clan – as well as the design of their clan's woven cloth and their clan's songs and dances. Members of the clan can be called upon to do voluntary work or to give livestock or money in lieu of voluntary work, when this is demanded of them. They also acknowledge traditional arbitration and justice systems for infringements of traditional laws and accept the authority of traditional leaders. The kind of CSOs with which CRS worked could have benefited greatly from collaboration with *Lia Nain*.

One aspect of traditional beliefs and structures came to the fore in the work to reconcile people who had supported the Indonesian militia with their original communities. Many of them had left the country with independence and now wanted to return. Many such people had been involved in extreme violence, but now they wanted to be accepted back into their communities. This was possible, but there were traditional ceremonies to undergo, often involving communal betel nut chewing and public apologies. There was something similar in Rwanda following the genocide called the Gacaca courts.

What did I learn from the coalface?

In post-disaster situations where there is a large influx of foreign resources, and these are accompanied by foreigners who are full of sympathy and pity for the plight of the locals, it is very easy to develop a "Santa Claus" mentality of both providing and expecting goodies. Foreign organisations with "tough love" are needed that will identify the capacities and capabilities of local people and help them develop their own abilities, rather than undercut these by paying too much attention to their victimhood.

There is likely to be a strong network of traditional ties between local people in any country where you work that is often not widely known. During 18 months' work in Timor-Leste, no foreigner ever talked to me about sacred houses and *Lia Nain*, nor did I ask Timorese friends about them until they volunteered information. They are, however, the real civil society of the country, and I wish I had learned much more about them when I first came to the country. In general, a lack of cultural awareness is an important deficit when trying to help people to help themselves.

It is very important to have clear, frank and pragmatic conversations between civil society and government in every country. I suggest that this cannot be left to individual biased and prejudiced people from each side. It is so important that it needs professional and trained intermediaries or counsellors. Countries can do themselves considerable harm by deepening and widening the gaps between the two blocks.

Development agencies also need to be very sensitive when setting up new structures which may clash with traditional structures.

What happened to it all?

Even after the violence of the period around independence and the need for outside peacekeeping troops was over, to my astonishment violence broke out all over again in 2006, after I had left. It seems to have been caused by grievances among soldiers of the F-FDTL (Timor-Leste

Defence Force) who belonged to different areas and who claimed that they were treated differently, and the struggle between political leaders in the run-up to the 2007 elections. The application of a relevant UN programme called DDRR (Disarmament, Demobilization, Rehabilitation and Reintegration) within the military would have been very valuable, but it was, I think, neglected.

Another issue was the vexed question of land. There was traditional ownership of land; there was land and housing that was taken over by the Indonesians, and whose ownership was unresolved; and there was land and housing belonging to people who had emigrated to Australia, Indonesian Timor or other places and who came back after independence wanting to claim their rights and property. There was tension between those who had stayed and fought and those who went into exile and then came back when it was all over. This was very similar to South Sudan.

The long-term economic independence and health of Timor-Leste is bound up with access to oil in the Timor Sea, which has been negotiated and renegotiated with Australia. Timor-Leste has set up a Sovereign Wealth Fund from its oil sales and has unexpected resources that we were not aware of when I was there.

There were many opportunities for insights into the situation of Timor-Leste. Here are some worth noting:

- The 2004 UEFA European Football Championship final between Portugal and Greece took place while I was in Dili and illustrated how people regarded Portugal. It took place at about two in the morning, and each time the Portuguese scored a goal you heard the cheers go up from the richer part of town and the groans from the poorer part of town – and vice versa. Greece won, and the next morning a few likely lads appeared in the streets wearing hastily constructed shirts in the Greek colours, and cheering Greece while despising Portugal. Portugal and the Portuguese were favoured by the rich and educated, but no one else, as far as I could see.
- The Portuguese-speaking senior-level officials of government – including many Timorese back from considerable periods in Mozambique – insisted that the law courts were conducted in Portuguese, using Portuguese laws. There were very few indigenous judges, and thus the courts were held before various short-term expatriate judges – Nigerians, Indians and others. The courts had to be conducted in English, Portuguese, Indonesian and Tetum with translation both ways. It was a mess.
- I met, serendipitously, with what have been called the Disaster Cowboys – people who made their living by serially working in

famous disasters: Cambodia, the former Yugoslavia, Afghanistan, Rwanda and now East Timor. Many of them knew each other and it was like an old boys' and girls' reunion. It was also interesting that some UN agencies brought locals with whom they had worked in the former Yugoslavia violence to be experts in Timor in security and police work. One subset of the Disaster Cowboys were the Disaster Researchers, whose job it was to try to make a true factual account of the violence that had occurred. Every day they counted records of bodies and atrocities – a strange way to make a living. A huge summing-up of the Indonesian era was produced in 2005 – the Report of the Truth and Reconciliation Commission – and, in 2008, the report of the Commission of Truth and Friendship was published, after a wait of 12 years.

- I also met serendipitously the commercial equivalent of the Disaster Cowboys – the UN Camp Followers. There were supermarkets for the kinds of goods foreigners wanted, bars and Thai massage parlours. It was really true that such establishments had been providing services to UN personnel on a regular basis in each place where there was a heavy UN presence. An added element in Dili was the presence of a load of Darwin and Northern Territory Australians who were skilled artisans – mechanics, air-conditioning, lighting, sewerage and building contractors. I was told that the word had gone out in Darwin, where many Timorese had gone to escape and find shelter, that there was work to be had in a pretty Wild West kind of place – i.e. Dili – and they had swarmed in. They were quite a riotous crew.
- A third significant group supporting the development of Timor-Leste was the solidarity organisations – from Australia, New Zealand, Ireland, the UK, Canada and the USA – that had been politically active in lobbying for East Timor independence at the UN or with their own governments, or that had lobbied against munitions going to Indonesia to attack East Timor (see my original contact through Oxfam with Timor when it was still Indonesian and I met Lord Carrington (Chapter 8)). Many people from such organisations came triumphantly to East Timor and were greatly welcomed for their work. Having arrived, however, they were at a loss – they were not development workers but political activists. Many of them got involved with the political struggles between East Timor and Australia over the ownership of the oil in the Timor Sea.
- Coffee had been cultivated in Timor-Leste from the early nineteenth century and some smart US agriculturists and marketing

experts pointed out that all East Timorese coffee was organic since very little pesticides or fertilisers had been used during the Indonesian period, and certainly not before. East Timor coffee could therefore command a premium price from environmentally conscious coffee importers as organic coffee, and it soon developed a reputation at Starbucks. Wild coffee grows in a singular way in Timor-Leste – in long, tall vines, which have to be pulled down to harvest the berries. It was starting to be cultivated on small shrubs by the time I left.

And on a personal note

While I was with the Partnership in Jakarta, my wife was working for UNICEF in Jakarta too, and my children were in boarding school in Singapore (at the very fine United World College). When I got the job in East Timor we tried to find a job for Clare in the same country but were unable to do so. We thus had a period of two years in which we were not living together – she was in Jakarta (and was lent to UNICEF Myanmar for six months) and I was in Dili. We would meet at intervals at the mid-point – Bali. This was tough (although no one could call Bali tough), but it was a pattern of life for many people in the international development business.

What were we thinking at the time?

It is not my memory that there was much comparative literature or research on Timor-Leste in 2002, apart from yards of project proposals and reports that normally stayed in the organisations producing them. What was available was:

Medan, Natasha. 2002. *From Resistance to Nation Building: The changing role of civil society in East Timor*. World Bank, Washington DC, USA.

Holloway, Richard. 2004. *What Is Civil Society in East Timor?* Self-published, available on Google.

Recently, more material has become available which was not available at the time I was there.

Wigglesworth, Ann. 2012. The growth of civil society in Timor-Leste: three moments of activism. *Journal of Contemporary Asia*, 43 (1): 51–74. https://doi.org/10.1080/00472336.2012.735545ADB. 2018. *Civil Society Organisations in Timor-Leste Brief*. Asian Development Bank (ADB), Manila, Philippines.

CAVR. 2013. *Chega! The final report of the Timor-Leste Commission for Reception, Truth and Reconciliation*. KPG in collaboration with STP/Comissão de Acolhimento, Verdade e Reconciliação de Timor Leste (CAVR), Dili, Timor-Leste.

Kammen, Douglas. 2019. *Independent Timor-Leste: Between coercion and consent*. Cambridge University Press, Cambridge, UK.

Notes

1 The programme was called Enhancing Civil Society Programmes in Timor-Leste (ECSP).

2 Pancasila was the national ideology of Indonesia, created after independence, and exported to East Timor as their thirty-third province.

3 One spin-off from this was that Indonesian was very widely spoken. This was linked to the 24-year occupation by Indonesians, the settling in Timor of many Indonesians as civil servants, and, in some cases, relocated peasants from other parts of Indonesia. I worked most of the time in Indonesian, occasionally needing translation into Tetum. The decision to make Portuguese the national language was initially divisive since it was very much seen as the language of the educated elite.

4 CRS. 2002. *CSOs in Timor-Leste*. Catholic Relief Services (CRS), Timor-Leste.

5 CRS. 2004. *Advocacy in Timor-Leste*. Catholic Relief Services (CRS), Timor-Leste.

15 Tajikistan, 2005–10

Persuading ex-apparatchiks that citizens can do good without the State

... In which the government officials inherited from the Communist era were persuaded of the importance of CSOs, which could do things that the government had given up on or was incompetent to do.

The Aga Khan Foundation (AKF) advertised in *The Economist* for a manager of their Civil Society Programme, and I found myself being interviewed by phone at midnight in Dili, East Timor, given the time difference between Geneva and East Timor. I got the job, and for the first time in my life started working out of Europe and being on the ground in serial visits rather than living in the country which was the focus of my work.

The first job was trying to understand what AKF understood by civil society. My predecessor had used the terminology of non-government organisations, but the new programme was conceptualised as promoting civil society as a necessary and important part of any society – not just to build and strengthen CSOs but also to build a civil society – and saw that in terms of government integrity and corporate social responsibility. Of course, each country in which I worked contextualised this differently, which was very exciting.

My job took me to eight countries including those I knew already (Kenya, Uganda, Tanzania) and others that were brand new for me (Mali, Pakistan, Afghanistan, Tajikistan, Kyrgyzstan). There were thus many trips to places I had never been before – especially in Central Asia. I had to learn from the CSOs that I worked with there what it meant to be at the grassroots or the coalface in their countries, particularly when we were encouraged to stay in the expensive hotels that were owned by the Aga Khan Development Network (AKDN)!

Working out of Geneva was also very different from working out of the towns I had been in before – Maun, Juba, Roseau, Lusaka or

Jakarta. Working for such a large organisation was also strange, and a first for me. Strangest of all was being invited to occasional board meetings of AKF in the Aga Khan's headquarters in Aiglemont, France, where dark suits were obligatory. The attraction of working for AKF was that it seemed very serious about sustainability and taking a long-term perspective on development initiatives. Its most famous investment was the AKRSP (Aga Khan Rural Support Programme) in North-West Pakistan, in the regions of Gilgit and Baltistan. It had been working there for over ten years and seemed ready to continue. This was such a difference from the usual three-year projects that depended on the fickle serial interest of a Western donor. Another attractive aspect was that AKF and AKDN were ready to put their own funds into projects they identified to supplement the funds they sought from national and international donors.

Once it was clarified with AKF what kinds of programmes and projects they were expecting from me, I had to raise the money, and in collaboration with AKF in the UK we made a proposal to the UK's DFID and received funding for a five-year programme. At last, a long-term time horizon for working with civil society.

This chapter is about working on civil society issues in Tajikistan, a country that emerged from the Soviet Union as one of the poorest of its previous members. It also has a large population of followers of the Aga Khan within its borders. AKF and its affiliated organisations were very important in rescuing the country from extreme poverty and famine at the end of the Soviet period and the civil war that ensued.

However, because I think that many readers will be interested in the singular nature of AKF, the singular development work it supported me to do in Tajikistan, and the whole issue of supporting CSOs in a post-Communist State, I will take some time to introduce the context.

To the general public who may have heard of him, the Aga Khan is known as a very rich person and perhaps as a racehorse owner: very few know that he is the 49th Imam of the Ismaili sect within Shia Islam (which, along with the Sunni, makes up the Islamic religion). Few also know that he has created a very large development foundation with the singular approach of allying business and commerce, culture and social development with not-for-profit development work. A lot of his work involves his relationship with his religious followers, scattered over about 20 countries, for whom he is their Imam. Apart from that he governs and manages the development agency called the Aga Khan Development Network, one part of which is the Aga Khan Foundation. AKF employed me to manage one of their six programmes: the Civil Society Programme.

I need to provide some clarity about the range and size of AKDN before going deeper into the issues concerned with the Foundation, civil society development and Tajikistan. AKF is a singular organisation, and when trying to explain it to others I have often used the idea of a cross between George Soros and his Open Society Foundations and the Dalai Lama with his followers in Tibetan Buddhism. AKDN has activities in 19 countries and incorporates work in economic, social and cultural development. These examples from Africa are likely to be the best known:

- Work in the field of **Economic Development** (such as Tourism Promotion Services, Industrial Promotion Services, Financial Services, Media Services and Aviation Services). Probably many will be surprised to find out that Kenya's Nation Media Group, Uganda's Monitor Publications, Jubilee Insurance and Diamond Trust Bank Group, plus all the Serena Hotels and Uganda Airlines all belong to AKDN.
- Work in the field of **Social Development** (such as Agency for Microfinance, Universities in Pakistan and Central Asia, Education Services, Health Services, Planning and Building Services, Academies and Humanitarian Assistance – plus AKF, which incorporates work in Rural Development, Health, Education and Civil Society). Aga Khan hospitals, schools and academies are well known across East Africa.
- Work in the field of **Culture** (such as the Award for Architecture, Historic Cities Programme, Music Initiative, Museums and Exhibitions). The restoration of the mud mosques of Mali in Djenné, Mopti and Timbuktu and of Stone Town in Zanzibar, for instance, has been carried out by AKDN.

It is astonishing to become aware, as you work for AKDN (even if only on one programme – Civil Society – within one agency, AKF) of the size and variety of AKDN's work. The Aga Khan Foundation was started in 1967 shortly after the Aga Khan graduated from Harvard University, and was conceptualised to be similar to the Ford Foundation. It is based in Geneva but has offices in all the countries where it has substantial programmes. Its literature claims that it gives special emphasis to the needs of rural communities in mountainous, coastal and other resource-poor areas. And the countries served have nearly all had some connection with the Ismailis as they have evolved and travelled over the last millennium.

One branch of Ismailis started in Fatimid Egypt in about 900 AD and moved out to Syria and Iran when faced with military aggression from the Mongols. Eventually the Khoja branch of the Ismailis moved to Pakistan, and from there to India. With guidance from past Imams (particularly the present Aga Khan's grandfather), they were encouraged to move to Africa, and so came to Kenya, Zanzibar and mainland Tanzania, Uganda, Mozambique and South Africa. Another branch of Ismailis, following the missionary activities of Nasir Khusraw, who was from Afghanistan but was converted to Ismailism on a visit to Cairo in Fatimid times, established Ismaili communities in Gorno-Badakhshan in Tajikistan; the Baghlan, Badakhshan and Bamiyan areas of Afghanistan; and the Gilgit/Baltistan areas of North-West Pakistan. Many of the services offered by AKDN were originally to meet the needs of the Ismaili communities in Asia and Africa, but today they have moved beyond that. AKDN programmes now work with non-Ismaili populations, but they also occur in areas where the Ismaili community is not present, e.g. Egypt, India, Kyrgyzstan and Mali.

The link between the Aga Khan and Tajikistan is also an interesting story. The sizeable population of Ismailis in Gorno-Badakhshan were forbidden to worship their Imam during the years of the Soviet Union, where Tajikistan was one of the satellite states. They also could not meet him, and contact with their Imam was thus blocked. At the end of the Soviet Union, the subsidies on which many of the Soviet satellite countries depended were cancelled and a civil war ensued in Tajikistan. The country was one of the most dependent Soviet satellites and was in a terrible state when aid was cut off. The Gorno-Badakhshan area, where the Ismailis lived, was in a worse state than most parts of the country. It was high, mountainous, and had only one road to the capital, Dushanbe, which was blocked by snow in the winter. I have travelled this road and it is daunting, snow covered or not. As soon as the borders were open Ismaili representatives from Tajikistan journeyed to Aiglemont, France and told the Aga Khan of their situation. This resulted in a huge relief operation of food, clothing and blankets as well as generators, vehicles and development programmes, which largely rescued his followers from starvation. AKDN and AKF then initiated a larger operation of development interventions for all of Tajikistan – banks, microfinance, rural development, education and industrial development, hydroelectric power – and one element of this was civil society.

All of this resulted in the Aga Khan Development Network and the Aga Khan Foundation having a singular position in Tajik society. One whole province of Tajikistan was recognised as being populated by the Aga Khan's followers. It was well recognised that AKDN and AKF had been very active in relief and were now active in development, and that both foreigners and locals were carrying out these activities. We were regarded with little suspicion. We were, however, operating in a country that had come out of the Soviet sphere of influence, and had to manoeuvre accordingly.

As part of the Soviet Union, Tajikistan had followed Russian norms, procedures, structures and language (although it also had its own language, Tajik). As with all ex-Soviet Union countries, their mindset was based on "Soviet heritage" – i.e. a system based on the directive role of the Communist Party, with the bodies of State authority duplicated by party authorities. In such Communist societies, which have now largely disappeared in this form (except in North Korea), people associated in bodies which were approved and managed by the Party (e.g. youth clubs and women's associations), and such bodies were not independent of the State. The Communist Party was the source of knowledge and authority. There were changes in the political system in Tajikistan from the first Constitution in 1994 to the approved amendments in the Constitution in 2003, which allowed CSOs to exist "as public associations for the protection of common interests, and the achievement of general purposes" and which established "the right of free access to existing public associations", which were defined as "voluntary, self-management and non-commercial formations created by citizens' and legal persons' initiative".

All of this suggests that the situation in Tajikistan was supportive of CSOs, at least in law, but the actual attitude of government officials was rather different. There were basically three kinds of public associations, and each of them was received differently by the old-style apparatchiks.

- Communist-era public associations were ones that were formed by and with the Communist Party: such organisations were accepted and established, but everyone knew that their resources had dried up and they might exist only in name. While they existed, however, they should not be challenged by others. They included Komsomol for youth, trade unions and cooperatives, as well as collective farms, creative associations – writers, artists, journalists – and women's associations.
- New, high-profile organisations were being created by international NGOs (Mercy Corps, Save the Children, ACTED, Oxfam) with foreign money (funding from USAID, Switzerland, the UK) as well as by UN agencies – especially UNDP, which tried to initiate *Jamoat*[1] Resource Centers. The government was suspicious of them and was worried that they might be anti-government.

- There were the traditional versions of voluntary self-organisation: joint neighbourhood committees for self-help (called *hasher* and *shariki*), community governance groups (*Jamoat* and *mahallas*) and religious organisations (*zakat*).
- Lastly, there was the whole range of new and small CSOs created by citizens aware that the legal framework allowed them to exist. Such bodies were in many cases formed because of the failure of the State to provide services and the need for local self-help groups to cope with the problems they had. In general, government officials were ignorant of them, dismissive of them, and suspicious that they might have some anti-government inclinations. In any case, officials assumed them to be small, insignificant and not worthy of their attention.

It is this last kind of organisation that AKF focused on. We tried to convince government officials that there were all sorts of citizens' organisations that existed at local government level, that they were doing good work and that government had no need to be wary of them: quite the contrary – that government should look at them as possible collaborators in working together to overcome problems that all knew to be real. There were extensive long dinners with vodka toasts to all and sundry, and powerful work by my Tajik staff to get me to understand the strange world in which we were working.

We were able to get the ear of Senior Adviser to the President, Isomiddin Sahahuddinov, and we provided information to him about what was being done at the grassroots, made a film in Tajik and English called *They Deserve to Be Recognized: Glimpses of the work of CSOs in Tajikistan*, and, with his help, and the help of CSO coordinating groups, arranged for senior government officials to visit the local authority areas to see what was being done. It was very successful. The majority of the organisations they visited were entirely Tajik – there was no involvement from foreigners.

We got reports back that the government officials had visited such organisations as:

- NGO "Fariza" in Khojent, which was involving local people in *zakat* philanthropy;
- Khujand Rehabilitation Centre, helping children disabled after the civil war;
- NGO "Nur" in Khorog, Gorno-Badakhshan, working with mentally challenged and deaf children;

- NGO "Volontyor" in Khorog, working with drug addicts and People Living with HIV/AIDS (PLWAs), providing clean needles – for many addicts needle cross-infection was the route by which they had got AIDS;
- NGO "Chilu Char Chasma" in Shartuuz, working with women returned from Afghanistan to make carpets, and then providing microfinance for them to start their own businesses;
- NGO "Internationalist" in Muminabad, Khatlon, working to provide clean drinking water;
- NGO "Ubed" in Khulob, offering vocational training in knitting and sewing for poor women; and
- "Fidokor" in Dushanbe, teaching CSOs about civic education.

The message came back to us from the senior government officials that they found dedicated, sincere local people working to improve their well-being and overcome the problems of the bad days of the civil war. They also found that local communities had already started to sit down with local government officials for the first time to solve common problems.

In private conversations, senior government officials talked emotionally about what they had seen in respect of disabled children and drug addicts – problems that they were very familiar with in their own lives.

Out of all this came a strong directive from the senior administration of the Tajikistan government that there should be regular round tables between government officials and CSOs at district level where they could explain to each other what they were doing and see if there were opportunities for collaborative and cooperative efforts. Mr Sahahuddinov recommended that there should be a programme of small grants from government offices to valuable CSOs. We heard no concerns about interfering foreigners or worrying Islamisation, but plenty about the value of self-help and the ways in which CSOs helped to bring this about.

Parallel to this was helping those larger and more educated CSOs which saw their role as trainers and teachers of CSOs to be better, more skilled and more professional in their work. It was always a tricky path to tread – to persuade government officials of the good work that was being done in the country entirely without their involvement. So many of them were still in the "Soviet heritage" mindset that all good things should come from government, even when it was clear in Tajikistan that the government was seriously poor and unable to provide the services that it had during the Soviet days.

What did I learn from the coalface?

Committed, active and dedicated people who are concerned to help their neighbours can usually convince sceptical and suspicious government officials that their motives are pure and wipe away concerns that they are anti-government, if government officials visit their work and do not rely on rumour.

It is very important to develop key contacts who will allow you to get your message to decision makers. It is difficult to arrange this but it comes from endless networking, and, in Tajikistan, with the trust that came from being a member of AKF.

In ex-Soviet countries there is a high level of educated, skilled and trained people in the country, even if it has gone through all manner of civil strife.

What happened to it all?

Seven or eight years after I left working with AKF, and working in Tajikistan, there was conflict between central government and the provincial Government of Gorno-Badakhshan which involved fighting between "commanders" – the name for veterans of the civil war who had kept their personal militias intact. There was even disagreement between the Ismailis in Khorog (the capital of Gorno-Badakhshan) about the strategic value of having their first allegiance to their Imam in a political situation in which they should be trying to get political power for themselves.

And on a personal note

At the end of my time at AKF I arranged to have a long trek with all my family and another close family down on the mountainous border between Tajikistan and Afghanistan and along the Silk Road, following the Amu River. From there we headed north, crossed into Kyrgyzstan, to Osh, then to the capital Bishkek through the high pastures, called *jailoo*.

AKF was interested in setting up a system of home stays on this route to bring another source of income to the people there, and we tested it out. It was great. It was also very interesting to see that the people walking in the mountains on the other side of the river in Afghanistan had fewer and poorer clothes and worse roads than on the Tajikistan side.

In the high *jailoo* of Kyrgyzstan, where we slept in yurts all the time, there were marmots which popped up all over the place to look at us. We were told that these animals were the source of the plague that came to

Europe in the form of the Black Death, and that they still have a reservoir of disease in their burrows.

What were we thinking at the time?

AKDN and AKF have a well-developed communications unit which makes available literature and films on its work in general and its work in the different countries where it is present. It also produces an in-house magazine, *The Ismaili,* which has features on its different countries of operation. This was all part of our daily fare when working at AKDN.

AKF. 2011. *AKF and AKDN: A continuum of development.* Aga Khan Foundation (AKF), Geneva, Switzerland. Available on the AKDN website at www.akdn.org (accessed 2018).

The AKDN website has a section on "Frequently asked questions" and there is also a website for "The Ismaili – Tajikistan" at https://the.ismaili/microsite/tajikistan (both accessed 2018).

And the AKDN-owned Institute of Ismaili Studies researched some of the more erudite information about the origins of Ismailis in Central Asia.

Niyozov, Sarfaroz. 2003. *Evolution of the Shi'a Ismaili Tradition in Central Asia*, Institute of Ismaili Studies, London, UK.

We produced our own material in Tajikistan and found some other international NGOs (INGOs) working in comparable fields. There have been very many INGOs working in Tajikistan subsequently and much more information about what was previously a very limited field.

Yusufbekov, Yusuff, Babajanov, Rustam and Kuntuvdiy, Natalya. 2007. *Civil Society Development in Tajikistan.* Allavida/AKDN, Dushanbe, Tajikistan and Farnham, UK.

There was also the film produced by AKDN in Tajikistan in 2007: *They Deserve to Be Recognized: Glimpses of the work on CSOs in Tajikistan.*

Note

1 *Jamoat* (or District) is the lowest unit of local government.

16 Different countries in Africa, 2005–10

Building integrity and CSO standards as an alternative to fighting corruption

... In which the governments and businesses of Kenya, Uganda and Tanzania were persuaded that CSOs were competent and needed development actors in their countries, and that building personal and organisational integrity was better than endlessly railing against corruption.

I had often felt that my work did not have enough pictures in it – not enough illustrations, cartoons and stories, and too much written material and workshop paraphernalia. The work with the Civil Society Programme (CSP) at the Aga Khan Foundation (AKF) enabled me to turn this corner – we produced over 30 comic books and seven educational and dramatic videos. And it was possible because of the huge range of interests and sectors represented by the Aga Khan Development Network (AKDN) and the funding from DFID.

Part of the objective of AKF's Civil Society Programme was to build a civil society in the countries where it operated – and this meant, in part, reducing corruption and building corporate social responsibility. AKDN writes, in answer to the FAQs on its website, that the project companies of AKFED[1] operate as businesses, but all surpluses are reinvested in further development initiatives. AKDN is also proud that its agencies are corruption-free, but it is very conscious that they often operate in countries of endemic corruption, and that there needs to be regular monitoring of the standards of the AKDN organisations.

The CSP linked up with two very interesting organisations that had a close fit with the aims and objectives of AKDN. These were the International Business Leaders Forum (previously the Prince of Wales International Business Leaders Forum) and Tiri (subsequently Integrity Action). With guidance from the Aga Khan, the CSP focused on public integrity as the coordinating principle of its work:

> A deeply rooted sense of public integrity means more than integrity in government, important as that must be. Ethical lapses in medicine and education, malfeasance in business and banking, dishonesty among journalists, scientists, engineers, or scholars – all of these weaknesses can undermine the most promising democracies.[2]

So the Institutional Integrity Programme was introduced within the CSP as a way of developing a civil society and of reducing corruption by increasing integrity. It used the examples of AKDN agencies to do this – in a range of sectors (hospitality, microfinance, rural development, secondary and tertiary education, health facilities and hospitals, the media and insurance) and from eight countries (Kenya, Uganda, Tanzania, Mali, Afghanistan, Pakistan, Tajikistan, Kyrgyzstan). It was recognised that, in general, AKDN entities were exemplars of good practice. Yet all with whom we worked informed us that no organisation working in an endemic context of poor governance and unethical practices could be complacent about challenges to their integrity. The related risks of contamination are very real.

At the same time, running activities that address corruption – however it is referred to – is also quite risky. As I have mentioned before, corruption is "an open secret". People know all about how it works but do not want to mention it in public, and certainly do not want to feel that they are being targeted or accused.[3]

We did three things in the Institutional Integrity Programme: we produced comic books with the objective of changing attitudes about integrity and used them in workshops with the staff of the participating organisations; we produced a manual (*Integrity Guidance Manual*) to help organisations reflect on their existing or planned activities from an integrity perspective, with the objective of institutionalising integrity and making it part of the "organisation's DNA"; and, lastly, and only in Kenya, we produced seven School Civics videos about corruption seen from a youth perspective, using the metaphor of sport.

Comic books[4]

Our first work was to produce comic books based on realistic cases, illustrating realistic working conditions for each of the organisations in each of the other sectors. These were in English, French, Swahili or Russian, depending on the countries where we worked. The "realistic cases" were developed in focus groups comprising the staff from these organisations. Once the broad outline of the comic books was agreed, they were finalised in workshops with the participating staff and were

Figure 16.1 Extract from a comic book on integrity dilemmas

very well received. The comic books provided questions at the end for the participants to ask themselves: "What would you have done at different stages of the story?" – thus allowing them to choose a particular track of professional or ethical behaviour (Figure 16.1).

Comic books defuse the tension that is often found in discussions of corruption and lapses of integrity. They allow people the freedom to identify with the characters in them and think freely about solutions to the problems identified. Their purpose is changing attitudes, not specifically changing behaviour at that time. Behaviour change was planned as a result of the training connected to the *Integrity Guidance Manual.*

Our thinking behind telling the stories of corruption by positing ethical dilemmas was based on my understanding and experience of how corruption worked in different societies. If you were faced with an ethical dilemma which offered you the need to pay (for instance) illegal levies demanded by teachers or school heads, or the chance to pay illegal bribes to get goods through customs quickly, then, if everyone around you was doing this, that is the path you were likely to follow. You might have voiced objections (to a greater or lesser degree), but you went ahead with the same behaviour as everyone else. Repetitions of such behaviour made it even more likely that you would ignore your objections and do the same as everyone else around you the next time the opportunity arose. If, on the other hand, you were offered the chance, through a story, to pause and reflect on the impact of your corrupt behaviour, and to decide, on that basis, to make a considered different choice, it illustrated to you that there were alternatives. The more times you decided to choose such alternatives, the more likely it was that this would be learned and practised behaviour, which might influence others.

Some examples of the comic books:

1. **Mali: *Silent Killer***
   This tells the story of Wandaplast, a profitable plastics company, but one with a problem with toxic waste. On the recommendation of the Chief Engineer Apoto they have disposed of harmful waste in the company's landfill site, but then squatters occupy that land and some have become sick.

   After the lively picture narrative has been shown, participants are asked to consider:

   - What ethical obligations, if any, does the company have to the affected squatters?
   - What ethical obligations, if any, did Chief Engineer Apoto have to these people?

2. **Uganda: *Lilina can't cook books?***[5]
   This tells the story of Lilina Kakooza, who is the Chief Accountant at Uwezo Ltd. Her older, long-time boss, Mr Kania, cannot produce proper receipts following an overseas trip and asks her to cover for him by cooking the books before an auditor's inspection. He launches a charm offensive. She is shattered that he expects her to ignore her personal integrity and her integrity to the company and do the dishonest thing he asks of her (Figure 16.1).

   This is a fast and comical narrative, but ends up by asking the participants:

   - Is it acceptable for Lilina to endanger management by refusing to cook the figures?
   - Should shareholders' confidence in Uwezo be protected by false figures?
   - Would you maintain a tough stand if a dishonest superior launched a charm offensive?
   - What would you do to preserve ethical behaviour without being rude?

3. **Kenya: *Alice and her reports***
Chief Editor Alice Mwihaki discusses with the proprietor of *The Daily Telegraph*, Patrick Ogwal, the accusation by others that it is being selective in its reporting of crimes by one political party because he is too close to the tribe involved. Alice stands strong on resisting political influence but does not know how many in the paper she can trust. Bricks are thrown through the windows of Patrick's house, warning him to stop reporting certain crimes, and he is threatened. His wife is very frightened. It looks like Patrick will submit to political pressure. In Kenya, this is a worryingly likely story. The comic is very lively and ends up by asking the participants:

- Should she resign? If she does, the newspaper's reputation will be damaged.
- Should she investigate who is behind the threats and expose them? Maybe the result will be that the threats are carried out.
- Should Alice try to reach out to external stakeholders to pressurise Patrick?

Developing stories like these with an organisation is an exciting exercise, as it gives the people involved the possibility of telling stories that illustrate the kind of world in which they live and work, which they have never been able to do before. The organisation that develops the story must be very careful to make sure that no one feels that they are being targeted by a real story. Providing people feel confident that a story is being invented, and yet gives them the chance to talk about the kinds of problems they have, the methodology works very well.

When the stories are presented in the workshop, using a projector that shows the pictures and the bubbles containing the main actors' speech, like a comic, all participants are encouraged to take on the roles of the characters in the story and read out their parts of the script. Because it is about a subject they know well, people are generally happy to act out and act up the stories. At the end of the story, when the questions are being posed, it needs a competent and professional workshop facilitator to help people focus on the ethical dilemmas, and particularly to ask people to go back in the story and say how they would have felt at particular points in the narrative.

The purpose of the comic books was to try to get an understanding of the ethical dilemmas that staff of the identified organisations might face. This involved getting people to be open about the problems they faced in their working lives, and also clarifying what integrity, values and ethical principles mean in an organisational environment.

Integrity Guidance Manual

Building awareness and changing attitudes are only means to the end of effecting changes about integrity. Change will be achieved when heightened awareness and operational practices are "institutionalised" in the form of day-to-day management. When "embedded" in this way, integrity will become part of the organisational DNA and will continue to be a sustainable part of the organisation's existence. The manual was intended to serve as a useful resource in raising the capacity of organisations to have integrity-based performance, and it defined three important elements of institutional integrity:

1. **Alignment**: This means consistency between what your organisation intends or says it is going to do and what it actually does.
2. **Competence**: This means your organisation's ability to meet performance objectives with a balanced mix of managerial and technical skills and attributes.
3. **Ethical behaviour**: This means your organisation's conformity with principles and commitments that have been established to inform its decision making and behaviour – based on a consciousness of what is legally, morally and professionally important, obligatory or permissible.

The manual introduces a very interesting and functional visual tool to help organisations assess their existing or planned actions to see whether they reveal a culture of integrity in the organisation. This is called the Integrity Prism (Figure 16.2). It suggests that the organisation looks at its actions/activities through a prism[6] that asks the following questions:

1. **Alignment**: Will/does this course of action support achievement of our vision, fulfilment of our missions and practice of our values?
2. **Competence**: Will/does this course of action demonstrate adherence to expected standards of performance quality?
3. **Ethical behaviour**: Will/does our conduct in pursuing this course of action meet expected standards of behaviour?

School Civics videos and integrity education for youth

We had put a lot of effort into institutionalising integrity in development agencies but asked ourselves whether we should also be addressing students – asking whether it is possible to prepare students to resist corruption in the world outside school by early exposure to integrity and ethical issues.

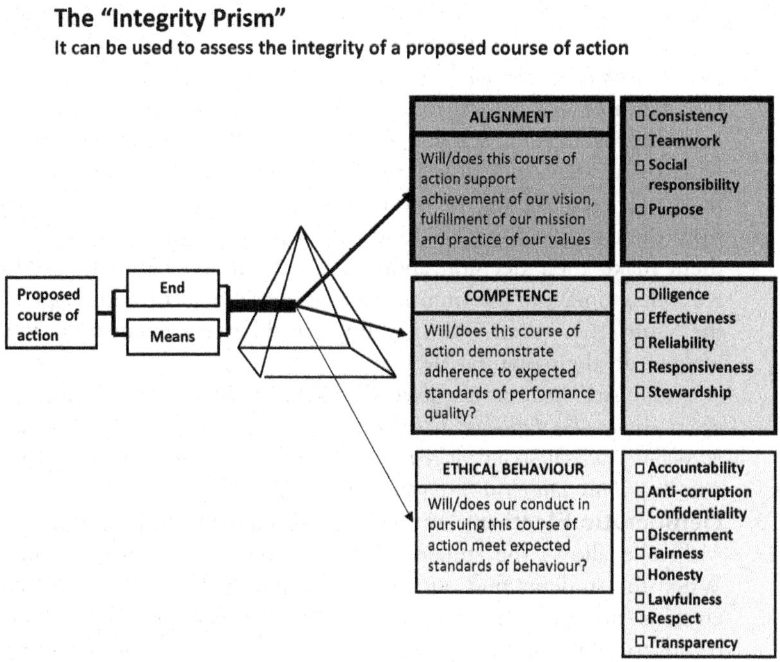

Figure 16.2 The Integrity Prism

We agreed to try to do this by posing corruption and integrity issues using the metaphor of sport and in the context of secondary schools in Kenya. We developed small dramas filmed with Kenyan youth and then suggested to the teachers to show the videos and discuss the issues that came up. At the end, the teachers would ask the students if they saw any examples outside school and in the real world which reflected these issues? Some very talented young actors and some very talented film editors produced these seven videos. Here are some examples:

1. **Democracy and Gender**: The girls' volleyball team and the boys' football team have both been selected to take part in international competitions. However, the school can afford to send only one team. The School Board decides to send the boys' team. The Headteacher wants to keep the Board's decision a secret until he announces the matter publicly. Unfortunately, the information is leaked to the female volleyball players, who protest that they are discriminated against because of their gender, not their athletic ability.

After considerable discussion about the contents of the video, the class is invited to contribute any examples they might know about in public life of organisations involved with poor governance.

2. **Governance**: Two local football clubs, Hearts FC and Rock FC, have approached Patrick and Owen from the senior class to play with them after they finish the school year. Patrick and Owen have heard conflicting opinions about the clubs and seek advice from their coach, Mr Keke, and the Deputy Headmistress to help them make their decision about which football club to join. Mr Keke has unfavourable information about one team and suggests they join the other. The Deputy Headmistress encourages the boys to do their own research.
   After considerable discussion about the video, the class is invited to think about people who influence important decisions and determine if there are possibilities of conflict of interest, and to brainstorm what unethical practices they think can occur in professional sport.
3. **Democratic Elections**: With the permission of the Headteacher, Mr Keke allows the students to elect their own games captain. When the students find out that the name of Owen, their captain elect, is not published on the student representatives' list, they begin to boycott the school until the responsible person they democratically elected is in office.
   After considerable discussion about the video, students are invited relate what happened in the film to the wider issue of how to ensure that the democratic process at community and national level can be free and fair.
4. **Conflict of Interest**: Mrs Tole, the girls' volleyball coach, and her husband are suffering from financial difficulties. Mrs Tole has been asked by the Headteacher to get new kit for the team. She decides to use her husband's business to supply her girls' volleyball team with new shoes for the upcoming season. However, because her family is in need of money, she asks her husband to find low-quality shoes for the girls so that they can keep the extra money in the budget for themselves.
   After considerable discussion about the video, the students are asked for examples of where the conflict of interest between the needs of the public and an individual's or small group of people's personal interests could have very serious consequences.

There are three more episodes: "Access to Information" about disputes in scoring in volleyball matches; "Judiciary and Neutrality" about handling a disputed decision about keeping score in a football match; and "Ethnic Diversity" about accusations that the coach is choosing someone from his ethnic group to play for the school.

I have never yet come across corruption and integrity education for young people (except perhaps in Hong Kong with the Hong Kong Police), although it is very common to hear comments along the lines of "This generation is lost to corruption – we have to try to make sure the next generation will not be the same." These videos are very educational – so much so that the Aga Khan Academy in Mombasa produced a pack for teachers to help them use the videos as an adjunct to teaching young people about democracy and ethics and to eschew corrupt self-interest.

What did I learn from the coalface?

- How important it is to learn the details of people's behaviour when discussing corruption – if you are describing the real activities of people's work experience, they will accept your interest and expertise in looking for ways to reform it.
- How important it is to avoid finger pointing and name calling when dealing with corruption – examples have to be taken from parallel or similar people and organisations which can illustrate comparative experience while avoiding targeting.
- Comics and cartoons are marvellous training tools. They can be used as flash cards, as cartoon storybooks or as videos with added dialogue. AKF and Integrity Action used comic book slides projected onto a screen and asked the audience to speak the dialogue out loud. It would also be possible to turn this into videos with dialogue, but we felt that participation was important.
- Educating youth about corruption, ethics and integrity is very important because, in its absence, young people will fall into endemic patterns of corruption – particularly if their parents ask them to give bribes on their behalf to teachers for higher marks or for promotion to higher classes.
- Using metaphor to teach integrity and contain corruption is a very interesting idea. We used the metaphor of sport with young people – taking examples from this limited universe and showing that this could reflect a larger context. The same could be done with many other limited universes – home life, looking after children, transport, etc.

What happened to it all?

- Unfortunately, neither Integrity Action nor AKF has continued to use the comic books. In my opinion, there is a lot of value in them.
- Similarly, the *Integrity Guidance Manual* is not being used further. It also has a lot of value.
- The School Civics videos are still part of the resources of Integrity Action and are used by the Aga Khan Academies.

Personnel recollections

The variety of stories that Tiri/Integrity Action developed with the staff of AKDN institutions was fascinating and obviously showed the reality behind people's working lives. The staff members of the organisations we worked with who had not been part of the creation/design process would exclaim with surprise when we showed them the comic books – it was very rare for people to have open discussions about corruption. When asked "What would you do?" at different stages of the story, people would suggest a very rich variety of alternatives. Discussions were very lively. As was mentioned in Chapter 13 on Indonesia, corruption is an open secret: people know all about it but don't like to talk about it.

What were we thinking at the time?

The creative process of working with the International Business Leaders Forum and Tiri/Integrity Action to develop these materials was very enjoyable. I believe these are very worthwhile learning materials and do not know many others like them. I am sorry they are not used more, but you can access the School Civics videos, the teachers' pack and the comic books on my website at www.richardholloway.org.

Tiri. 2009. *Integrity Guidance Manual*. Available at www.richardholloway.org/non-print-materials.

AKDN. 2010. Integrity education: a pack for teachers. Available at www.richardholloway.org/non-print-materials.

Tiri. 2008. Cartoon workbooks for the Institutional Integrity Programme. Available at www.richardholloway.org/non-print-materials/cartoon-workbooks-for-ethical-dilemmas.

Integrity Action has numerous case studies on the use of integrity on its website at www.integrityaction.org.

Notes

1 The Aga Khan Fund for Economic Development (AKFED) is the only development agency in AKDN that works on a commercial basis.
2 Aga Khan, graduation speech, Columbia University, 2006.
3 Robert Klitgaard's fine book *Controlling Corruption* narrates his tactic of describing corrupt practices in another country when trying to get a discussion on corruption started. Klitgaard, Robert. 1988. *Controlling Corruption*. University of California Press, Berkeley and Los Angeles, USA.
4 We were puzzled about how to refer to these "comic books": they told stories to adults with a message about integrity and were more akin to the long-established French tradition of the *bande dessinée*.
5 There is a pun here: "cooking the books" means reporting accounts dishonestly, and in the story Mr Kania asks Lilina if she is good at cooking.
6 The analogy is with a prism breaking up a beam of white light into its component colours.

17 Nepal, 2010–13

The birth of social accountability, digging down into corruption and half-hearted efforts to control it

… In which the World Bank introduced social accountability[1] *to Nepal through CSOs – how this new methodology was understood and how it was applied, but how complexities emerged when the associated corruption was managed by local government.*

On the DFID-supported project in the Aga Khan Foundation (AKF), there had been a worrying change in the exchange rate between pounds sterling and Swiss francs in 2008, which meant that we did not have sufficient money previously budgeted to finish the project. DFID accepted that they would let us have what was required, but this should come in the form of a supplementary project for which we needed to make a new proposal. Just at this time, Andrew Mitchell was appointed as the new head of DFID and insisted that all DFID projects comply with a new "value for money" mantra. All impacts should be shown numerically, with detailed information on how many people had benefited from each activity.

This was very difficult to graft onto a project whose mission was "to strengthen civil society organisations and build a civil society", but, if we wanted the money, we had to submit to the new conditions. A lot of international NGOs that were receiving DFID money were unhappy with these developments and there was a revolt that called itself "The Big Push Forward", but it got nowhere at the time.[2] The craziness of doing development by numbers slowly ameliorated, but DFID continued with its decision to favour large for-profit entities – many, like KPMG,[3] moved from being auditors to being money managers for development projects. It seemed to me to follow the path of USAID, which contracted "Beltway Bandits" (see Chapter 13) because "they knew how to manage money" rather than because they knew how to do development.

Halfway through this one-year extension to the DFID/Aga Khan Civil Society Programme, I was approached, to my surprise, by the

World Bank. They offered me the task of being the Coordinator of their programme to introduce "social accountability" to Nepal, which involved going to live in Kathmandu. The World Bank had been responsible for researching and operationalising the concept of DFGG (Demand for Good Governance) in its projects and had initiated its first project in Cambodia under the title of PECSA (Programme to Enhance Capacity for Social Accountability).

The thinking was now more solid, and I was asked to take it over in Nepal, with the title of PRAN (Program for Accountability in Nepal). Normally I would keep a long way away from World Bank projects, which nearly always involved working with government ministries that I could not trust, but I was assured that I would be able to work with CSOs primarily, which would be trying to hold government accountable.

"Social accountability" was to become a major strand in my life from this point on, and I need to explain it. It is not helped by a name which seems to come from a sociology PhD. What it basically means is the citizen being able to hold government accountable for public benefits that have been passed by legal, policy or budget directives *but which have not actually happened or been implemented.* Frequently, it is defined and described by public administration types who talk about "helping government to effectively tackle accountability problems, or deal with maladministration and error", but my experience is that it boils down, almost inevitability, to corruption – to government officials finding ways of escaping doing what they should be doing (and what they know they should be doing) in order to make illegal income for themselves.

In Nepal, we found four national CSOs to work with. They, in turn, found 32 smaller district-level CSOs to administer the project. These CSOs were well acquainted with government's (especially local government's) intentional bad management so that local government officials could get extra income, and they educated me on how this all worked within the local administrations of Nepal.

Nepal had just come out of ten years of civil war between the monarchy and the Maoists, usually called the Maoist Insurgency. The rebellion had been launched by the Communist Party of Nepal – Maoist (CPN-M) on 13 February 1996 with the main aim of overthrowing the Nepalese monarchy and establishing a People's Republic. It ended with the comprehensive Peace Accord signed on 21 November 2006. Many outsiders were surprised to find people in 2010 who called themselves Maoists since China's Cultural Revolution was long over and China had adopted its own form of capitalism. Nepal's Maoist insurgency was, however, very real. Wikipedia says:

> The conflict was characterized by summary executions, massacres, purges, kidnapping and other war crimes and crimes against humanity. The revolution resulted in deaths of over 17,000 people involving civilians, insurgents, army and police personnel, and internal displacement of hundreds of thousands of people (mostly of rural Nepal).

The Comprehensive Peace Accord proclaimed a whole raft of new laws and rights that had been part of the Maoists' platform, and the government machinery was duty bound to enforce them. However, many local government officials had been, and were, far away from the Maoist rebellion. They were not inspired to make the new laws work for the benefit of the poor and powerless. Moreover, Nepal had deeply embedded cultural patterns of behaviour that still oppressed substantial numbers of its poorer citizens – particularly Dalits[4] and women.

So I asked my CSO friends to introduce me to some of the wrongs that were not being righted as my first introduction to social accountability. "Go to some remote villages and ask the old people about their pensions," I was advised. So I tried. Most villages are remote in Nepal: field visits involve a lot of walking – or praying, if you are in a shaky vehicle on a narrow mountain road. I was helped to meet a group of elders in a remote village, and I asked them about their pensions.

"What do you mean, pensions?" they asked. They did not know that they were entitled to pensions. This was not by chance. The government officials at the district level knew that the elderly were entitled to pensions, but they had never passed on that information, let alone the actual pension money, so that they could keep that part of the district government budget for themselves. Therefore the locals did not know to ask for their pensions.

We explained what pensions were, how much they entailed and how they should be distributed. "Good Lord," they said. "We have been cheated! How do we follow up on this?" We told them the address of the District Council and the names of the relevant officers, and they determined to undertake the several days' walk from the mountain villages to the district capital. I was not present when they met the government officials, but I was told by the CSO friend that they were informed there was no longer any money for their pensions because it had been spent on a primary school. They were familiar with this excuse[5] and knew that the money had been stolen from them, but they did not know what they could do about it.

The government officials had broken the following laws and regulations:

- They had not informed the citizens of their rights and entitlements.
- They had not given them their rights and entitlements.
- They had changed the use of the elements of a local government budget.
- And they had not offered to help the citizens reclaim their rights.

This was the field in which PRAN would operate – social accountability. The usual way to try to address issues of social accountability was by investigating what was actually happening, holding open meetings in which this could be discussed, and then putting in place reforms and, if necessary, restitutions. The instruments were often public hearings, public (or social) audits, community scorecards and citizens' report cards.

After some time designing the approach, we put the PRAN programme to work.

As I have said before, the Government of Nepal had passed, in Parliament and subsequently in the Constitutional Assembly, some very fine laws, policies, statutes and regulations to govern the social contract. Indeed, if all of these were implemented, there would be little need for social accountability. The problem is that they were not implemented, and, as a result, WPEs (women, poor and excluded)[6] did not receive the services that they should by right. Moreover, they did not know what their rights or entitlements actually were. The result of this, in turn, was that they did not ask for them.

A lot of the activities of the CSOs we worked with were thus concerned with making citizens' rights and entitlements known to the citizens, particularly the WPEs, and then helping them gain access to these. This breakdown in the social contract rested on top of (and was possibly caused by) a society that was and is deeply riven by caste, class and ethnicity. Those in government responsible for upholding the rights and proving entitlements to the citizenry are largely educated upper-caste men who are not immediately motivated to help uneducated lower-caste women and other marginalised people, even though the law not only says that they can but actively encourages them to do so.

And in a layer on top of this again, the political situation of local government in 2011 specifically amplified this:

- Elections for People's Representatives for local government had not been held for 12 years.

- A hybrid formation called the All Party Mechanism had been put in place to replace this vacuum, until it was declared absolutely corrupt by the CIAA[7] and disbanded in July 2012.
- Corruption at national and local levels was and is endemic and practised with impunity. This had been comprehensively researched.[8]

PRAN found that it was possible, though difficult, to inform WPEs of their rights and entitlements, inform and educate government officials on what they should already know about these rights and entitlements, and bring these together in a constructive engagement. Our hope was to get government officials to recognise that the citizens had the right to know about and demand their entitlements, and get them to be prepared to accommodate these demands.

It is important to be clear that social accountability was not a new concept to the Government of Nepal. The government had incorporated public hearings and public audits into its procedures and had made them mandatory for infrastructure projects. The government also had extensive instructions for involving citizens in participatory planning and participatory budgeting of local government through the Good Governance Act. However, as the sub-grantees of the PRAN project started sending in their periodic reports, we found that these instructions were often ignored or hijacked by political parties, something that was noted in a report from the Asia Foundation. Sometimes local government was aware of and appreciated what it had not done and was prepared to improve the situation.

What usually happened was that the CSO staff made a courtesy visit to the offices of local government, informed them what they intended to do, and got their agreement to it in principle. In some cases the CSO was already well known to the local government because there had been a history of collaboration: e.g. VDRC (Vijaya Development Resource Centre) in Nawalparasi helped government schools with libraries. In other cases the CSO had to introduce itself as well as the planned project. In many cases the initial agreement from the local government was not an issue, but it became an issue later when the local government found out that the CSO's implementation of social accountability tools exposed their wrongdoing and caused them embarrassment. The next stage was usually that the CSO introduced the concept of social accountability to the citizenry of whatever area they were covering. This was novel since most CSOs were known for their service delivery. Social accountability did not mean delivering new services, but getting the government to deliver services which they should already have delivered. This stage involved a considerable amount of discussion and the holding of meetings. The CSO often held public meetings at which it urged citizens to ask for accountability from government agencies – something that few citizens were used to doing and were often apprehensive about doing.

The next stage was the collection of data and the public production of a report to local government agencies which showed what had not been done and should have been done, followed by discussions to try to get commitment from the local government to right the wrongs of the past. In quite a few cases the local government bodies were not aware of the laws and regulations that bound them and were surprised about what they had to do. In other cases they denied wrongdoing until the facts were incontrovertible. In some cases both parties came to a consensus about what needed to be done; in other cases it was more confrontational, as, for example, in the work of Sahara Nepal in Bajhang (see later), where popular pressure resulted in the repayment of 240,000 Nepali rupees that had been wrongly taken from the block grant for women's empowerment and, according to the VDC (Village Development Council), used to build schools, something doubted by the citizens.

What does this illustrate? That CSOs, trained by PRAN through its partners, were quite capable of learning the use of social accountability tools, of negotiating with local government for the implementation of these tools, of educating and orienting citizens about their rights and entitlements, and of openly practising using these tools. It could be that they were able to make the whole exercise positive and constructive, increasing the practice of good governance at the local government level in Nepal, or it could be that the exercise increased distrust and disaffection between citizens and local government.

It is instructive to recognise what did *not* happen:

- Local government never forbad the CSOs from practising social accountability (although there were instances where local government officials were less than enthusiastic).
- The CSOs were able to work with citizens at their own pace and in their own language. There were no instances where the citizens rejected the approach of the CSOs.
- There were no instances of confrontation, although sometimes the CSOs had to be skilful negotiators to avoid this, particularly with forthright groups of Dalits and women.

In most cases the project led to a commitment from local government officials to reform their ways of working so that they kept to the law and the regulations in respect of rights and entitlements, and to a commitment from citizens to regularly monitor the local government officials in the future to make sure that there was no backsliding and that they would take up their legally approved opportunities for participatory planning.

It was interesting, however, to see the behaviour of the citizens who had taken part in these exercises and, by doing so, had revealed injustices. Even though many of the activities of the PRAN sub-grantees involved identifying and controlling corruption based on individuals and political parties taking for themselves resources that rightfully belonged to poor and marginalised people like themselves, it seemed that few citizens and citizen groups were interested in sanctions, punishment or prosecution of the wrongdoers. Apart from Sahara Nepal in Bajhang, there was no evidence over two years of PRAN's work of citizens demanding restitution for money stolen from them, or a desire to take the cases to the courts – perhaps because they felt that they had no likelihood of restitution from the courts.

Citizens took a different approach. I particularly remember a women's group in Pokhara that had found out that the money budgeted for their training had been used for something else (inevitably a primary school). Instead of confrontation, they told me that they fully intended to inform the local government officials that:

- they now knew what their rights and entitlements were;
- they now knew that they had been robbed of their rights and entitlements; and
- they did not intend to let this happen again.

They were putting the local government officials on notice that they would be watching for future infringements. I asked them if they did not want to try to get their money back and they politely but firmly told me that this was their choice of action.

When we looked back at what we learned from two years of practising social accountability through PRAN, the following points came up regularly in the reports of sub-grantees:

- More time was needed to make a serious change in the status quo. There was a change in the mindset and then a change in practices.
- Many people, particularly the very poor, needed allowances to allow them to take part in voluntary activities, particularly discussions (which were important parts of any social accountability work), as

this took them away from earning their daily income. Going to meetings in Nepal required considerable time for travel.

- Success came from reforms carefully negotiated with local government officials.
- It took more time than had been envisioned to teach local people that social accountability was a useful method of work to improve their lives.

Once taught, however, people were very enthusiastic and committed to continue the work. I have mentioned Sahara Nepal as one of the few groups that contested the stealing and managed to get their money back. This is their story:

Sahara Nepal

The Village Development Councils (VDCs) of Kanda and Rilu in the District of Bajhang (along with the rest of Nepal) had not had local councillors elected since 2001. In the absence of elected councillors, the VDC secretaries acted as both government officials managing local government and representatives of the citizens. They were, however, usually members of the elite and did not represent the citizens. They were, in turn, "advised" by the members of the All Party Mechanism (APM) which represented political parties and was formed, similarly, from village elites. While all APMs had been formally abolished in 2012 because of their contribution to corruption, their actors and activities continued.

Nepal's budget provided resources to the poor and marginalised, and, on paper, there was an adequate participatory local governance system to check on their expenditure. However, this often did not work in reality. Representatives of the poor and marginalised, while they had the right to attend the meetings of local governance bodies, were rarely involved in decision making and, most importantly, did not know the composition of the budget or how it was spent. PRAN trained a local CSO, Sahara Nepal, in participatory budget monitoring and public expenditure tracking so that they, in turn, could educate the poor and marginalised about the VDC budget and their part in it.

Sahara Nepal sought to discover the VDC budget from the VDC secretary and was only able to do so by threatening to use the Right to Information Act. When the secretary saw that they were serious, he capitulated and provided the information. Once they knew the budget allocations, they systematically informed the citizens of Rilu and Kanda of these and quickly discovered that allocations for women's empowerment (192,000 rupees), senior citizens and the disabled (72,000 rupees) had not been spent on these topics but illegally transferred to budgets for infrastructure and administration. The citizens angrily confronted the VDC with the facts and demanded that these funds be restored to their proper purpose.

Sahara Nepal mediated between the two parties but also told the whole story to Saipal FM, a local community radio station that broadcast it throughout the Bajhang district. They also reported the situation to the Local Development Officer (LDO). The LDO sent a circular letter to government departments that such behaviour should cease, and Sahara Nepal was able to persuade the relevant government officials to return the funds to their intended purpose. The local citizens determined that they would use their legal rights and processes to obtain their due entitlements in the future.

In most cases, the national-level CSOs taught the local-level CSOs and community-based organisations (CBOs) various kinds of social accountability tools. The most common tools in order of frequency of their use[9] were:

- public expenditure tracking systems – systematically tracking what money had been used for;
- participatory budgeting – involving citizens in planning budgets for public services;
- citizens' charters – drawing up a publicly displayed list of the tasks of a government office and what should be paid for each task;
- public hearings – public meetings between citizens and government officials to clarify specific problems; and
- community scorecards – a way for citizens and service providers to identify the most important issues in a chosen field (e.g. schooling), to score their importance to them, and then to negotiate a plan to improve the situation.

In order to help the national-level CSOs with their work, PRAN contracted a local media organisation to make video training films for the different tools; these were then used in workshops to train citizens. The videos included different media – live footage, cartoons and simulations – and were found to be very popular. The basis for the training videos were the following questions:

- What is meant by this accountability tool?
- Why is it important?
- What can go wrong in using it?
- What can citizens do in such cases?

The videos were in Nepali and were filmed within the Nepali culture. They had subtitles in English, and I have used them in many different countries. The basis of social accountability (poor and marginalised people trying to get accountability from richer and more powerful people) is understood everywhere. What was new was planning for "Plan B" at the start – trying to think in advance about how your strategy might be blocked.

What did I learn from the coalface?

- In spite of situations where villagers had had funds stolen from them, there were very few instances where villages wanted to confront government officials. The usual tactic was to make sure that the government officials knew that the villagers knew what they had done and that they would not keep quiet next time.
- When given the opportunity to do role plays or simulations, villagers were very accurate in portraying the illegal activities of the government officials.
- Villagers had a strong level of trust that senior government officials would understand their problems and would side with them against local government officials.
- In Nepal, corruption for personal income was less common than corruption for income for your political party (from which, of course, you would get some benefit).
- It takes two years for a social accountability programme dealing with government budgets to come to fruition. In the first year the corruption is researched and exposed, and the second year provides an opportunity for people to practise some countermeasures to stop it.

- Although corruption in Nepal, and particularly party political corruption, was an open secret, very few donors were prepared to address it. The three pieces of research that dealt with corruption at the grassroots are listed in note 8.

What happened to it all?

- The World Bank's PRAN was the first donor-funded social accountability programme in Nepal. It was followed by ActionAid, Integrity Action and Partnership Transparency Fund, and a few more.
- The complexities of political rivalry and political parties continued long after I left Nepal. This remained the basis for continued corruption, as political parties had no other source of income.
- A massive earthquake destroyed many parts of Nepal in 2014. There was a huge international response with funding, but because the corrupt practices were still endemic in the country, a lot of the aid was stolen.

Looking back from a personal perspective, Nepal, like Ethiopia, is an amazing place for trekking. My children and their friends – all university students by this time and pretty fit – trekked to Everest Base Camp and were astonished to discover that a quarter of their number had to go back downhill fast because of altitude sickness.

My wife and I, and one daughter, with guides, travelled north-east of Kathmandu to see the huge indigenous rhododendron forests. They were fabulous, but the trekking was also very hard work – and risky. On one path we followed, the mountainside just in front of us split off and crashed like a locomotive into the valley below. It was a near thing.

Part of the joy of Kathmandu is the grey-haired hippies left over from the hippie trail of the Sixties and Seventies. They make a thin living by offering bed and breakfast and by growing and selling organic food. They are easily outnumbered by the aid workers, and by a new breed, the voluntourists, who come to work in orphanages or with children. While I was there, a researcher held a workshop on the history of the hippies – they had become the subject of sociological research.

But forget the foreigners – Nepal is a staggeringly beautiful place and its people are very interesting, including retired Gurkhas who have built houses for themselves that look like Aldershot, and old sadhus who just look out of it all the time.

What were we thinking at the time?

The World Bank had "discovered" social accountability in 2004 and had written many sourcebooks on the subject – and continues to do so. Its first deliberate project was in Cambodia, and this was well written up. What was interesting in Nepal was to discover the close correlation between social accountability and corruption and the ways in which the funds for public services for citizens were disguised and used for other purposes. As I have already said, there were good laws and directives in place for social accountability measures to be applied by government offices: the problem was that they were not well implemented.

We added to the literature on social accountability and introduced the training videos:

Holloway, Richard. 2013. *Social Accountability in Action: Illustrations of PRAN's work in Nepal 2009–2012*. World Bank and PRAN, Kathmandu, Nepal.

Khadka, Kedar and Bhatterai, Chianjibi. 2013. *Source Book for 21 Social Accountability Tools*. World Bank and PRAN, Kathmandu, Nepal.

In 2010, PRAN Nepal produced 11 15-minute animated cartoons in Nepali with English subtitles (available at www.richardholloway.org/non-print-materials/social-accountability-training-videos):

- Social Accountability Tools (general)
- Public Hearings
- Citizens Charters
- Right to Information
- Public Expenditure Tracking
- Budgets of Local Bodies
- Public Audits
- Community Scorecards
- Civic Education
- Checklists of Standards and Indicators
- Checklists of Entitlements.

These added to the useful materials on social accountability from the World Bank, ActionAid, UNDP and others:

World Bank. 2006. *Social Accountability Sourcebook*. World Bank, Washington DC, USA.

Malena, Carmen, Forster, Reiner and Singh, Janmejay. 2004. *Social Accountability: An introduction to the concept and emerging practice*. World Bank Social Development Papers, Washington DC, USA.

World Bank. 2010. *Opening the Black Box: Contextual drivers of social accountability effectiveness*. World Bank, Washington DC, USA.

UNDP. 2008. *Fostering Social Accountability: From principle to practice: guidance*. UNDP, Oslo, Norway.

In ActionAid's "Just and Democratic Local Governance" series:

ActionAid. 2011. *Accountability: Quality and equity in public service provision*. ActionAid International Governance Team (no place of publication).

ActionAid. 2011. *Budgets: Revenues and financing public service provision*. ActionAid International Governance Team.

ActionAid. 2012. *Democracy: Justice and accountability at the local level*. ActionAid International Governance Team.

ActionAid. 2012. *Power: Elite capture and hidden influence*. ActionAid International Governance Team.

ActionAid. 2010. *Using Evidence to Establish Accountability: A sourcebook on democratic accountability for development practitioners and learning facilitators*. MS ActionAid-Denmark, Copenhagen, Denmark.

The World Bank has subsequently enlarged its interest in social accountability by initiating a programme called ANSA (Affiliated Networks in Social Accountability) and GPSA (Global Partnership in Social Accountability).

Notes

1 Social accountability is, unfortunately, difficult to explain. Try "Why is the government not doing what has been agreed by law? And what can we do to make this happen?"

2 The Big Push Forward conference was "trying to reconcile messy, unpredictable and risky pathways of societal transformation with bureaucracy driven protocols" (September 2016).

3 KPMG is one of the Big Four accounting firms.

4 "In Nepal political and economic power was consolidated by interlinking it with the Hindu caste system" (World Bank and DFID. 2006. *Unequal Citizens: Gender, caste and ethnic exclusion in Nepal*. World Bank and DFID Nepal, Kathmandu, Nepal). Dalits were the caste, formerly called "untouchables", that undertook "impure" occupations.

5 The way this worked was as follows: a local authority would be responsible for building needed local infrastructure, such as local schools or clinics. They would issue calls for tenders and decide which bids would be awarded the contracts. Although there were rules for such practices, it was well known that bids were inflated and awarded to contractor friends of the local administration, who, by this method, received extra income and paid kickbacks to their friends in local government.

6 This was the terminology of the World Bank in Nepal at this time.

7 The Commission for the Investigation of the Abuse of Authority (CIAA) was a government body.

8 There are a number of reference books on corruption in Nepal: Asia Foundation. 2014. *Political Economy Analysis of Local Governance in Nepal.* Asia Foundation, Kathmandu, Nepal; Dix, Sarah. 2011. *Corruption and Anti-corruption in Nepal: Lessons learned and possible future initiatives.* Norwegian Agency for Development Cooperation (Norad), Oslo, Norway; and the National Planning Commission's study on corruption in social security payments: National Planning Commission. 2012. *Assessment of Social Security Allowance Program in Nepal,* Government of Nepal, Kathmandu, Nepal.

9 Written instructions on how to use social accountability tools can be found in: PRAN. 2011. *Sourcebook of 21 Social Accountability Tools in Nepal.* PRAN, Kathmandu, Nepal; World Bank. 1995. *Sourcebook of Social Accountability.* World Bank, Washington DC, USA; UNDP. 2008. *Handbook on Social Accountability.* UNDP, New York, USA.

18 Myanmar, 2015–16

Watching a country become aid-dependent and doing nothing about corruption

... In which the brave new world of Myanmar lost its gloss for development workers and I worked with a local CSO on an EU project to reduce illegal logging – and what this revealed about corruption in Myanmar is still an unaddressed subject.

After I left Nepal, there was no long-term work on offer and so I looked for short-term consultancies. These are very varied. In some cases they are very exciting, as you deal with a new and innovative idea: but at other times they are very boring – you are much more like a bean counter, as you try to check how many indicators among some donor's requirements have been achieved. Whatever kind of consultancy such jobs are, it is a frustrating way of working because they usually last three to five weeks, so you are not around long enough to see any results.

The first consultancy in Indonesia – which was indeed very interesting – was to evaluate a project called *Program Representasi* which was intended to help improve the extent to which Indonesian parliamentarians represented their constituencies. The norm in Indonesia was for an MP to occupy a parliamentary seat as a sinecure, or as a means to promote business interests, with little contact between them and the people who had voted for them. We discovered many of both kinds of parliamentarians and there was evidence that the project had had some effect in getting MPs to be more accountable to their constituencies.

The second consultancy – even more interesting – was to evaluate for DFID a Norwegian organisation's project in social accountability in Rwanda (PPIMA – Public Policy Information, Monitoring and Advocacy). Rwanda is a most complex country. Following the genocide, the President initiated a centrally controlled and very innovative development regime for the country, part of which prohibited corruption. There was, it is true, little opportunity for a democratic opposition and plenty of evidence of a top-down way of working through government ministries. I also saw a lot of evidence in the villages of citizens being given the opportunity to discuss development alternatives. Where citizens found that the government offices were not doing what they should be doing, there were clear opportunities to report this behaviour up through the hierarchy. There was also, it seemed to me, plenty of opportunity for Rwandan CSOs to exist and to be recognised by the government as valuable development partners. My contribution to the evaluation suggested that PPIMA was suitably rounding out citizens' ability to hold local government to account and, through their work on advocacy, enhancing citizens' ability to hold national government accountable too. Donors, however, were very focused on alleged human rights abuses by the government and, to my way of thinking, did not pay enough attention to what CSOs and community-based organisations (CBOs) were able to do in the country.

The third consultancy, for the World Bank, was much less interesting and valuable. Following its start-up work in social accountability, the World Bank had set up a network of organisations practising social accountability around the world which they called the "Affiliated Networks for Social Accountability" (ANSAs) and for which they supplied the initial three-year funding. They set up ANSAs for sub-Saharan Africa, South Asia and South East Asia, and, most recently, the Arab Region, building on the events of the Arab Spring. This was the one they wanted me to evaluate. This network was based in CARE in Egypt and operated in seven countries of the Arab world, and I was asked to evaluate it with an Egyptian NGO. I found that the World Bank had cut the funds in Year 3 and that the other ANSAs had collapsed or were collapsing. There was not much to learn or report from this consultancy.

So, after three consultancies, I was keen to get back into the field. Where were the most exciting development opportunities in 2014? For me it was clear: Myanmar! This bizarre country had been ruled erratically by military dictators since the end of the Second World War, had the longest continuous civil war between its lowland core people and its ethnic armed groups, and, until Hurricane Nargis, had not allowed foreign NGOs to enter the country. It was poised for extraordinary elections in which the woman who had been under the government's house arrest for 15 years was contesting as the leader of a new political party, the government was building a completely new capital city in the countryside 200 miles away from Rangoon, and there was a surge of tricky foreign investment for resource extraction, legal and illegal. Myanmar promised to be a lively place.

My previous experience of Myanmar had been while it was still Burma.[1] In 1987 I had been asked by CUSO (see Chapter 9) to visit Myanmar to examine the possibility of them opening up a programme there. At that time, foreigners were allowed visas for only a week's visit for a pre-planned official tourist itinerary. CUSO had discovered that a professor from McGill University in Montreal had two PhD students at Yezin Agricultural University in Myanmar for whom he was planning a supervision visit. He agreed to apply for a visa for me as his assistant and get me to meet the Principal to ask about placing agricultural *coopérants* at his university, or in the country. He advised disguising the purpose of the interview from potential onlookers, and so, knowing the keen interest of senior Myanmar government officials in golf, we carried DVDs of golf instruction to present to him as a gift. The discussions about possible Canadian volunteers came to nothing. A hefty cash investment would be needed to interest the government. I subsequently journeyed on to what was then Rangoon and visited the orthopaedic hospital pretty well run by the Burmese military to meet the needs of soldiers injured in the war with armed ethnic groups. I found Swiss volunteers from the International Committee of the Red Cross (ICRC) based in Geneva setting up the manufacture of the Jaipur foot[2] prostheses.

SPECTRUM and ECCDI

Unfortunately, in 2014 I was unable to persuade any of the usual development agencies to employ me in Myanmar, and so I looked at the possibilities of being a volunteer – something I had last done in Ethiopia in 1966! VSO (Voluntary Service Overseas), the British volunteer sending organisation, offered me a job to do capacity-building training for Myanmar CSOs for a year within a British Council programme. I accepted with alacrity, although the disparity between a World Bank salary and my VSO volunteer allowance was something shocking. Apart from the specifics of what I might do with these organisations, I wanted to be in Myanmar and to be part of the evolution of that enigmatic country. The government accepted only volunteers for CSOs that were registered with them. This forced me to work with two organisations: one legal and registered, and the other less legal and less registered. So I started work with two CSOs, called ECCDI (working on environmental concerns) and SPECTRUM (working on alternative development strategies).

While I was working with these two organisations, the preparations for the elections were heating up and everyone was on tenterhooks, trying to guess what the existing regime would allow and what would be their sticking points about further democratisation in the country. It was a time of great optimism, which was, unfortunately, overenthusiastic. The background to the country during the election excitement was unfortunately no different from what it had been previously – rich and rapacious government/military crony businesses acquiring land and other natural resources illegally, trading in narcotics, evicting farmers from their land and dispossessing others, plus continuing military operations against what were called the "ethnic armed groups" (i.e. the minority hill tribes and nations). Not much had changed although everyone was hopeful. Everyone expected that things would get better once the new political party, the National League for Democracy (NLD), took power and once the Lady (the name everyone used for Aung San Suu Kyi) was in charge – but everyone was also anxious that the military would not let this happen.

Living in run-down Yangon (as Rangoon is now called) was rough – many houses had not been painted since the end of the Second World War. I have never seen such astonishing spools of spaghetti-like electricity cables festooning the streets and the houses. The buses, which were all I could afford, were packed and dirty. They were so crowded that getting on and off at the right place involved intense rugby-like scrums. The food was astonishingly varied, using a great variety of (to me) little-known ingredients (kaffir lime, pickled tea leaves, fermented beans?). The language was really difficult. One of the great novelties was the strong connection all people had to their traditional clothing, be it in poor or rich fabrics. There was an elaborate system of knots to stop a man's sarong from falling down in extremis – much safer than the equivalent in Thailand or Indonesia.

One of the activities I undertook in this year educated but also saddened me about growing donor dependency. I undertook to give a training course for the CSOs, supported by the British Council, in fundraising and financial sustainability, quite similar to what I described in Chapter 12. This taught me a lot about the country. Buddhist Myanmar headed the list of the most philanthropic countries in the world from 2014 to 2017;[3] one might assume, therefore, that they had less need of foreign aid. I soon found out that people were not thinking like that.

Buddhism puts a great emphasis on donations. Monks collect these from house to house, or on the street, daily. Much of the income from these donations pays for the costs of the religious institutions and the upkeep of the monks and nuns, but Buddhist temples are also responsible for many educational, welfare, religious and cultural activities. Monasteries are also involved in responding to natural disasters, large and small. There was, however, very little philanthropic giving to Myanmar CSOs. My feeling is that: (1) charitable CSO were largely unknown if they were not connected to Buddhist temples; and (2) those that existed were not yet trusted by the citizenry. For these reasons, and also because foreign donor organisations that had arrived since Hurricane Nargis promoted themselves and their aid, Myanmar CSOs increasingly looked to foreign organisations to support them.

There was, almost by definition, no support of Myanmar CSOs by the government, unless you falsely accepted the government-run and -owned youth, women's and farmers' organisations as independent CSOs. There was a small but growing amount of business philanthropy to CSOs by businesses that were keen to "clean their faces", but, as I was soon told in no uncertain terms, such businesses were the "crony capitalists" that were supported by and supporters of the hated military government. Many CSOs had strong antipathy for the "cronies", who they saw as being responsible for environmental and livelihood destruction in the country, and would not take money from them even if it was offered.

So we were sadly left with a situation common to so many countries: local CSOs were dependent on foreign aid donors, not just for their money but also for the activities and projects that the money could be spent on. Myanmar had had the real opportunity of avoiding this, being a country so temperamentally inclined to personal philanthropy, but it did not do so and is now as donor-dependent as so many other countries.

ALARM

After a year working with the two Myanmar CSOs, principally helping them to write proposals to foreign donors to support their work, I circulated my CV around organisations that I respected and asked if they were interested in hiring me. I thought that a year's experience on the ground in Myanmar would be a selling point. I was contacted by a very interesting Myanmar CSO called ALARM whose objective was better forest governance. They had received a three-year grant from the EU to help them combat illegal logging. This project had the elaborate name "Forest Law Enforcement, Governance and Trade, Myanmar – Laying Foundations and Mobilising Civil Society" (FLEGT). It proclaimed that it used a multi-stakeholder approach, meaning that the CSOs would collaborate with government and business in order to reduce illegal logging. The objective was to achieve a VPA (Voluntary Partnership Agreement) by which all three sectors would agree on a legal framework for reducing illegal logging and for not selling illegal logs (particularly the famous Burmese teak) outside the country.

What sounded admirable proved to be complex, hypocritical and corrupt. Corrupt elements of the Forestry Department conspired with the Myanmar Army to fell timber illegally and sell it over the border to China. The ethnic armed groups (Karen, Kachin and others) also felled illegal timber and sold it over the same border, sometimes on their own account and sometimes in cahoots with the army and the government. Sometimes Chinese commercial companies came over the border with all the heavy equipment needed, paid off the ethnic armed groups and/or the Myanmar military, and/or the Forestry Department, and/or the border guards, and took the timber back over the border with them. Burmese teak was so valuable and so attractive that the public governance of the forests to preserve teak for future generations was subverted in every case.

In my opinion there never was a possibility of achieving such a VPA. Both government (through the Ministry of Environment and Forestry) and the timber companies were up to their necks in illegal logging, although loudly proclaiming that this was not so, or proclaiming that while this was a little bit so, they were doing everything they could to stop it. The Forestry Department's strong card was that they said that they alone claimed the right to determine if timber had been felled illegally or not, and thus whether it could be legally exported. They claimed that they could track the felled timber they wanted to sell back to the site of its felling and show that it was felled legally. This referred to the timber that was openly traded – to Europe, the USA and Japan – and which had to have government certification. They were exposed, however, as fudging this certification, initially by Sweden, who called their bluff, and then by other teak-importing countries. Irrespective of this, however, everyone knew that there was a huge amount of teak being sold illegally over the border into India, China and Thailand, which no one even tried to certify. In some cases it was re-exported to Europe from those countries.

Our job at ALARM was to work with local CSOs to prove the felling was illegal and to provide such incontrovertible proof that the lies, hypocrisy and corruption of the government position were exposed – thus requiring reforms before a VPA could be negotiated. We hired forest researchers from Copenhagen and East Anglia Universities to do this research, printed and disseminated it, and held workshops about the situation. ALARM was enormously assisted by the highly skilled Environmental Investigation Agency (EIA) of the UK, which carried out in-depth undercover investigations of who was – or was not – doing what to whom, and what they claimed they were doing (or not doing). This had two results:

- It exposed the convoluted channels of corrupt felling and selling: under-scrutiny, over-extraction, under-counting, bribery of guards and bribery of officials for reclassification.
- It worked with European countries to remind them that it was their responsibility to check that the timber they imported was legal. It then proved to them that the certificates of legality provided by the Myanmar Timber Department were false and that they should not allow such certificated timber to be imported into their countries.

This was powerful stuff. It operated at many different levels. ALARM assisted by mobilising young activists to go into the woods and document illegal logging. This was dangerous, particularly if they disturbed illegal felling camps at work. ALARM worked to get local government to lend them forest guards to help such groups. But it also operated at very high-level symposia, where, in my opinion, untrustworthy European timber-purchasing firms squirmed at the exposure and so doubled up on their declarations of probity and ethical behaviour.

There were many countries ostensibly planning for a VPA – Malaysia had been planning for one for decades, which suggests how seriously they approached it. One of the Asian countries that had succeeded was Indonesia, and the FLEGT project agreed to have a study tour of how government, businesses and CSOs were able to work together convincingly there. I was enrolled as the Indonesian–English interpreter. It certainly seemed that Indonesia had a working set-up – the Indonesian CSOs (some of which I knew from the past) were serious people, and the Myanmar participants on the study tour were impressed.

The problem for ALARM, and for so many sincere organisations working for the future of Myanmar, was that all the big guys had impunity. There would be no one taken to court; even if that did happen, no one was convicted unless it was approved by the highest reaches of the NLD and the army because their own people were the guilty ones.

There was also the kind of schizophrenia I saw at the Forestry Department. Officials there were proud of their heritage and their professionalism under the British colonial administration; they would then decry the damage that was done (against their wishes) by the Generals, who had rapaciously demanded that they deliver unreplaceable amounts of timber to the regime; they would then declare their desire for reform and improvement of the system, all the while knowing that the illegal extraction was continuing. I saw this in the field of forestry but believe that something similar exists in mineral extraction, land clearing for agriculture, industrial investment and many other areas.

As part of the work with ALARM I got to know a big timber processor who made all sorts of products from teak. He was well aware of all the corruption in the Forestry Department – he had five ministry guards at his factory whom he regularly paid to look the other way – and while he was willing to talk about it in Burmese with fellow citizens, he was unable to be open about it with me in English, although he spoke English perfectly well. He used all manner of circumlocutions, paraphrases and soft synonyms for corruption, rolled his eyes and gesticulated. We both knew what we were talking about but a certain kind of schizophrenia made him unable to vocalise it formally. Technically, he was well aware that teak was being exhausted and he had started laying thin teak veneers onto other timbers.

At the time of the elections there had been huge optimism about all the ways in which the NLD would be able to stop the excesses of the past, bring development for the future and promote democracy. But almost before the elections were over, sharp analysts saw that Aung San Suu Kyi and her party had very different ideas. I was there but was not experienced enough in the country to see the signs. Political conversations also took place in the Myanmar language and I was not part of them.

These were the signs:

- Before the elections, aggressive, political and popular Buddhist monks who belonged to a pressure group called MaBaTha loudly and vehemently denounced Muslims. In response, Aung San Suu Kyi did not have any Muslim MPs in the NLD structure.
- Before Aung San Suu Kyi was put under house arrest, when she was campaigning, and when there was a huge amount of political violence against activists for the NLD, many of her followers were arrested and given long sentences in harsh conditions. In 2014, when the Lady was released, they were also released and called themselves the 88 Generation of political prisoners, and they became a powerful pool of activists. They had suffered great punishment and they wanted reform. However, Aung San Suu Kyi had no office holders from the 88 Generation in the NLD.

- Substantial numbers of educated Burmese had left Burma during the reign of the Generals but had come back after 2010. Many of these formed CSOs; again, they were not invited into the NLD.

In contrast, the NLD seemed to be cosying up to "cronies", Burma's tycoon class that enriched themselves and their military patrons once authoritarian capitalism become not just allowed but the norm. The "cronies" donated hugely to the coffers of the NLD.

Through my work with ALARM, I was very well informed about corruption in the timber trade. Further conversations with Myanmar citizens and CSO activists about public services provided for the citizenry showed how the country was polarising into those who could afford expensive private services and those – the poor – who put up with increasingly underfunded and poor-quality public services. The fewer the public services, the more opportunities there were for public officials to demand bribes for access to such limited services – teachers, doctors, vets, water and electricity workers.

One deeply embedded corrupt practice was that teachers insisted that students take extra tuition with them, paid for by their parents, who could not afford it. There were very real threats that students' exam results would be penalised unless this happened. I was told that there was a high-level attempt to stop this practice but the teachers rebelled, saying that their government wages were insufficient to live on and that they had to have this extorted supplement.

All purveyors of the aid trade were aware that they were pumping money into a country whose governance practices were corrupt. Therefore, there was a very high chance of money being diverted away from the purposes for which it had been given. And yet there was little interest by the donors in researching or documenting corruption. Corruption stories were seen as anecdotal, impressionistic and difficult to prove. ALARM had proven them in the field of illegal logging, but no donors seemed to want to do this in any other field.[4]

Private conversations with donors at the time I lived there were along the lines of "Give the Lady a chance, cut her some slack, give her an opportunity to improve things before we attack her and her government on corruption issues." To me this is unacceptable – many donors had an anti-poverty underpinning to their aid, and the poor are clearly the worst hit by corruption in government because they cannot access the services that government should be providing for them. Since then the "halo" has been removed from Aung San Suu Kyi because of the Rohingya genocide, and it may be that more donors are not prepared to give her any more slack and will be more concerned with endemic and harmful corruption.[5]

What did I learn from the coalface?

- Young activists require protection in situations where they are exposing illegal activities. Law enforcement agencies need to be engaged on their side, but this may be tricky if these agencies have already been contracted by the opposition.
- Donors are hypocritical in the way their rhetoric is at odds with the knowledge that they have. This should be exposed, and those who operate ethically should be praised and promoted. "Diplomatic language" is, all too frequently, a politically correct cover for lies.
- Getting to the bottom of what are the actual drivers of endemic and established corruption is no small task. It takes a lot of research, people and time, and donors have to consider this to be worthwhile.
- Those who have been political prisoners and have lived through long and cruel imprisonment, and who are ready to keep their cause alive when they are freed, are amazing and praiseworthy people.

What happened to it all?

- The big issue is the Rohingya genocide and ethnic cleansing, which illustrates, if it is necessary to do so, the hypocrisy, denial and lying that are a standard tactic of the military in Myanmar. That has always been their modus operandi – it has not changed and it must stop. What has changed, as everyone knows, is that the Lady, who was so honoured and so praised, is now part of that corrupt way of operating. She has lost her honour and her reputation. I do not know what this means from the perspective of the realpolitik between the Myanmar Government and foreign donors, but presumably a great deal more scepticism. I discovered that DFID had refused to continue funding the FLEGT process.
- When I was there with ALARM, the Myanmar Army was not a well-liked, trusted or approved organisation. In ALARM, there were many staff from the ethnic groups that had supported armed groups against the army. There were also many lowland "Burman", who had the same opinion of the army. What was extraordinary to me was that, once the Rohingya crisis broke out, people were prepared to praise the army for defending the homeland and for persecuting the Muslims, whom, I found to my shock, were greatly vilified by large numbers of Buddhist Myanmar citizens.

And on a personal note

My wife, Clare, was in France while I was in Myanmar but came out on a long trip in which we followed up on the astonishing story of George Orwell (Eric Blair). Clare had spent four months in Burma in 2002 for UNICEF and, as a family, we had all made several tourist trips to Burma before then. Blair's Burma story is amazing when you appreciate what he did later in life, including fighting in the Spanish Civil War and writing *Nineteen Eighty-Four.*

George Orwell came from a British/Burmese family and once he left school he returned as a military policeman in the colonial administration of Burma. He wrote a powerful book called *Burmese Days* about that experience, based on the time when he lived in a town called Kathar (not the name in the book), a port on the Irrawaddy about 12 hours by boat north of Mandalay. With the help of local enthusiasts who could see some niche tourism possibilities in George Orwell, we visited his old house (now occupied by a Myanmar police official, suitably), the old British Club with its tennis court (now occupied by a cooperative), and a building that has been set up as a kind of George Orwell museum.

What were we thinking at the time?

ALARM, with EU support, was researching and exposing illegal logging, helped greatly by the EIA, and we were using their research for our work. The nature of corruption in the timber trade seemed to me to be clear, but no one from the government or the donors was interested in prosecuting.

Springate-Baginski, Oliver, Treue, Thorsten and Htun, Kyaw. 2016. *Legally and Illegally Logged Out: The status of Myanmar's timber sector and options for reform*. EcoDev and ALARM, Yangon, Myanmar.

EIA. 2019. *State of Corruption: Crime and corruption in Myanmar's teak trade*. Environmental Investigation Agency (EIA), London, UK.

I tried, unsuccessfully, to interest donors in pursuing corruption issues, using ALARM's work on timber as a model.

Holloway, Richard. 2016. *Corruption: The open secret of Myanmar*. Self-published.

I also tried to educate myself about Myanmar through the few foreign anthropologists and sociologists working on the country. There was much material available in the Myanmar language, but I could not read it.

Prasse-Freeman, Elliott. 2016. Burma's historic elections. *Anthropology Today*, 32 (1): 3–4. https://doi.org/10.1111/1467-8322.12222.

Prasse-Freeman, Elliott and Latt, Phyo Win. 2018. Class and inequality in contemporary Myanmar. In Simpson, Adam, Farrelly, Nicholas and Holliday, Ian (eds), *Routledge Handbook of Contemporary Myanmar.* Routledge, Abingdon, UK and New York, USA.

Notes

1 Burma became Myanmar in 1989. The government claimed that "Burma" referred to the lowland "Bamar" majority, but "Myanmar" incorporated all the people of the country. This, however, is challenged: the consensus is that "Myanmar" is a formal literary designation and "Burma" more everyday.

2 The Jaipur foot is a tremendous piece of appropriate technology from India. It is a lower-leg prosthesis for people who squat to defecate. It is made of recycled auto tyres (newer versions use plastic tubing) and has a flexible ankle joint. Its use has restored millions of disabled people's mobility in Asia.

3 CAF (Charities Aid Foundation) World Giving Index 2018. By 2018 it had fallen to seventh place.

4 The only people who were researching corruption were those connected to corporate ethical behaviour, particularly the Myanmar Centre for Responsible Business.

5 See Prasse-Freeman and Latt on "Class and inequality in contemporary Myanmar".

19 East Africa, 2018–19

Social accountability neutered by corruption

... In which eager, enthusiastic local national volunteers enlightened citizens in these countries about their rights and entitlements – and discovered how difficult it was, and still is, to overcome entrenched and institutionalised corruption.

After I came back from Myanmar I did a couple of consultancies from home while looking for something longer term overseas, preferably in the field of social accountability, which now seemed to me to be the sharpest knife in the drawer.

The British organisation VSO (Voluntary Service Overseas) put out an advertisement which attracted me a great deal. They wanted someone to train their staff, volunteers and partners in the meaning and techniques of social accountability in Kenya, Tanzania and Uganda. As an organisation they had decided to be more "rights-based" and less "service delivery-based", and for them this translated into three approaches to all their work: Social Accountability, Gender and Social Inclusion, and Resilience.

Because of my previous knowledge of VSO, I was apprehensive about what they meant by "volunteers". I was not enthusiastic about training Europeans who might be in the country for only a year and would have little understanding of local accountability. However, and to my delight, in these three countries of East Africa, volunteers meant "national volunteers": that is, local young people who were keen to get involved in field-level development work in their own countries. What they meant by "partners" was local CSOs from these countries who were interested in the same topics that VSO was. And VSO's fields of work were education, livelihoods and health.

The job would mean being based in Nairobi, Kenya, with repeated visits of approximately ten days to the other countries. I had been in all of these countries while working for the Aga Khan Foundation, and

Nairobi in particular had been my "Kansas City" (by that I mean the end of a long trek) when I used to drive there overland from South Sudan in the 1970s (see Chapter 3).

Social accountability (reprise)

If we accept the concept of a social contract to explain the relationship between those governed and those governing, then we accept that citizenship in a country (often signified by people's readiness to pay taxes) gives citizens to certain rights and entitlements, usually laid out in constitutions and laws. These rights and entitlements have been expanded by countries signing international conventions and agreements, and then "domesticating" these international conventions into their own national laws.

Social accountability suggests that if these rights and entitlements exist in law, then the citizens of that country should receive the services set out in these laws. Where this is not happening, they should be able to hold the State accountable for these laws, which, in effect, the State has promised to observe. Different countries, depending on the devolution or decentralisation of their governments, will offer citizens different ways of holding government accountable at different levels. The basic point, however, is that citizens, particularly the poor and marginalised, have the overriding right to hold the government accountable for what has been promised them in its constitutions and laws. Taking this down to lower levels, citizens have the legal right to be involved in the development of plans and budgets that affect what the local government does and how it does it.

Social accountability practice suggests that communities who are trying, at different levels, to hold government to account for implementation of their rights and entitlements should try:

- to identify a particular issue which is commonly felt to be important in that community;
- to aggregate their numbers so that government can see that the issue has a wide ownership (and is not important only to a few); and
- to make sure of information and objective data so that communities can refer to the actual laws and information which are the basis of the points they wish to make, and to the reality of the problems they face.

This job gave me the opportunity to train cadres of young people who were active in their communities to introduce the principles – and, hopefully, the practices – of social accountability. My introduction was often:

"We are not talking about getting more money for development, we are talking about what has happened to the money already agreed." There would usually be a pause while this was digested, and then a delighted roar of approval.

My audiences for this training were young but experienced grassroots workers who understood how local government worked and how the poor and marginalised interacted with it. They had, however, a generalised and imprecise understanding of how poor people were not receiving what they were due and what they were entitled to, coupled with an emotional but imprecise feeling that somehow poor people were being cheated.

The workshops and training courses often turned out to have two parts: the first was on governance – clarifying citizens' rights under the various constitutions and thus what they could legitimately ask for; the second was on public services – to address their feeling that patterns of corrupt behaviour were so deeply rooted and endemic in the country that the application of social accountability would not make enough of a difference to distorted public services. They commonly felt that the diversion and maladministration of public services and funds were such that even the suggestion of social accountability practices was unlikely to work to the benefit of the poor and marginalised.

Governance

In all three countries the Constitution allowed citizens to be consulted on what they wanted to happen in their country. There were consultative opportunities at village and district level for citizens to be involved in governance. They could contribute to plans and request responses from the government structure. This was best known in Kenya because the new Constitution and devolution law were only 12 years old and so were still quite fresh in people's minds.[1]

Our workshops would often follow a path like this:

> "Do you know that you are allowed (indeed encouraged) to attend formal meetings and put your points of view to the local government structure?"
>
> "Yes."
>
> "Have you ever attended such meetings and put your point of view?"
>
> "No."
>
> "Why not?"
>
> "We are never told when such meetings are taking place, and when we discover after the event, they are often far away from where we live, and they take place in English, and they require us to read a lot of material (in English) if we are going to take part."

The structure was there, but the structure did not work for poor and remote people. I asked a government official in a remote place why it was that he advertised participatory government meetings in a newspaper that was read only in the capital city. He replied that this was indeed a problem and he wished he could overcome it. Was he consciously and intentionally disbarring people from such meetings? My feeling was that the causes were poor administration and lack of vision – he simply had not thought through what needed to be done to include poor and remote people, and he did not think it was important enough to worry about. And no one supervised him to check what was going on – or not going on.

In Uganda, the equivalent of Kenya's devolution law of 2012 had happened 30 years previously. Uganda had been lauded at that time for a well-designed participatory governance system, but when I talked to CSO staff involved in local governance issues, they laughed and said that it was old history. Everyone had forgotten about it, and government certainly was not emphasising it. A further point made by fieldworkers in Uganda was that, if citizens at the lowest levels had successfully put forward their suggestions for development planning, there was no certainty that these ideas would be agreed higher up the hierarchy. Often, what was agreed and budgeted would be very different from what was proposed. After a number of iterations of this, citizens lose confidence in public participation and stop attending the theoretically very valuable opportunities for them to put forward their views.

In Tanzania, people were certainly aware of their rights as citizens to put their points of view to government, but repeated experience of these views being ignored had led them to give up on these avenues for citizens' voices to be heard.

Finally, in Kenya, in one rural community I found someone who had attended one of the citizens' planning meetings. I asked her what had been her experience. She said that the meeting was dominated by people loudly looking for political patronage and favours from the government officials; this prevented anything useful from being discussed. And the higher officials "gamed" the system, following the letter but not the spirit of the law. One egregious example was the Governor of Machakos, one of Kenya's counties, who had taken many of his senior officials and businessmen to Dubai to look for possible investment.[2] While there he circulated a notice of a public meeting about Machakos's plans to his fellow travellers, held the meeting and got the approval he needed. He had thus ticked the box – public meeting held, Constitution followed – but there was no vestige of popular participation.

As well as getting feedback from the workshops, VSO undertook a survey into the opportunities for social accountability in nine counties of Kenya.[3] This study and its findings were discussed in the two other countries for comparison, and the fieldworkers there recognised that Kenya had much in common with their own countries. The main findings were that local politics (usually expressed as looking for opportunities for personal income) outweighed opportunities for citizen engagement.

Public services

Apart from the formal opportunities for citizens' voices to be heard (which were poorly used), people could use other means to try to point out to government officials when public services were not working, or were not being delivered to the required and agreed standards. In village *barazas* (town hall-type meetings), in personal meetings with government officials and even on community radio programmes, they would ask, in different ways, for this situation to be reformed and improved. To many people these informal processes seemed realistic and pragmatic, but experienced fieldworkers were pessimistic about such methods working. Their understanding of the reality of pervasive corruption supported their pessimism. Here are some examples:[4]

- **Education (local primary schools)**: The funds from the Ministry of Education to the school (capitation fees) are siphoned off at different levels on the journey between capital city and primary school, so the school is nearly always short of its required funding; school textbooks do not arrive in the numbers planned, as they are sold commercially; teachers are often absent from their classes, seeking to earn income from other sources to augment their low school salaries; teachers, including the headteacher, impose extra (illegal) levies on parents, which they have to pay against their children's inclusion in the school; and teachers often insist that students take extra (paid) lessons outside school with them.
- **Health (local health clinics)**: The drugs supplied by the Ministry of Health may not arrive in the quantities or be of the quality expected, as they are likely to be siphoned off by nurses and doctors and sold commercially in nearby pharmacies; nurses and doctors may be absent from their work because they are operating commercially to augment their income; and nurses may take bribes to ensure priority treatment of patients in the clinics.

- **Local infrastructure**: Officers in local government are responsible for designing, tendering and receiving bids, contracting, inspecting and signing off on schools, teachers' quarters, clinics, roads, bridges, culverts, markets and other structures, but the practices are often corrupt. In many cases the contracts are awarded to friends or relatives of local government personnel on the basis of bids which are not scrutinised properly, or the buildings are not inspected properly or are agreed to be finished once bribes are paid, often when the construction is inadequate.
- **Media**: Newspapers, community radio stations and social media are considered likely avenues for seeking information on citizens' rights and entitlements, reporting on their non-implementation and encouraging citizens to speak up on topics they consider important. There has been, however, a serious and increasing amount of authoritarian control of the media in all three countries recently. Informed fieldworkers are apprehensive about what they are legally allowed to say and do and what punishment they risk if they use the media in ways unacceptable to governments.
- **Power politics**: Here I am not referring to party politics but administrative and cultural barriers that disempower citizens and prevent their voices being heard. Time and time again the field staff participants said that poor, uneducated and marginalised people would find it impossible to implement successfully what was being suggested by social accountability strategies. The reality of many poor people's lives is that: (1) they do not know their rights; (2) they do not know how to acquire such rights, even if they do know about them; and (3) they do not know how to complain if they find that their rights are ignored or violated. Moreover, when faced with government executive officials or elected government officials, they lack the clout and power to disagree with them and to counter those who oppose their ideas of seeking accountability. They are also apprehensive that any opposition from them will result in punishment or they will find themselves prevented from receiving the desired public services.

One aspect of this is that many poor and uneducated people do not consider government services to be their right and entitlement but something that is at the personal discretion of government officials, who have to be pleaded with and begged or bribed to deliver the services that people should be getting by right.[5]

In some cases, the pressure that citizens feel whereby they must obey local authority figures and keep their opinions quiet is reinforced by cultural and traditional structures and systems. Although it is changing, women, particularly in Muslim societies, will not contradict the opinions of their husbands; village meetings will not listen to the opinions of women; young people's voices are not considered worth listening to in public meetings; and, in particular, the voices of physically and mentally disabled people are given little attention.

Successes of social accountability

There were enough examples of positive impact from social accountability interventions in all three countries for it to be worthwhile continuing and expanding their application. In Kenya, for instance, in the county of Laikipia, it was found by CSO workers that the amount spent on pre-primary education was miniscule, even though a larger amount had been budgeted in line with national guidelines. Pre-primary schools were inspected and shown to be very substandard. In particular, the photos of collapsing toilets, which scared children away from using them, provided a strong argument. A combination of publicly displaying the vastly underexpended budget, public meetings to complain about poor facilities and veiled threats to report the county's practices to the local parliament and the Ministry of Education resulted in a revised and improved budget allocation that citizens monitored.

Furthermore, all the countries of East Africa have very large youthful populations: over 60% of Kenyans, for instance, are under 25 years old. They understand poorly the governance systems of their own country, although these are taught in primary and secondary schools as "Civic Education", and young people are already disengaged, disgruntled and effectively disenfranchised. Any other civic education takes place at the time of elections and is often very heavily politicised, providing polarised information to party political supporters' groups.

Maladministration or corruption

Maladministration

It is often said that poor public service delivery derives from government officials' lack of knowledge and understanding of the complex rules and regulations that govern, for instance, education, health, agricultural extension and social welfare provisions. Not only do the intended recipients of government services not know what they should be receiving, and thus cannot accurately ask for them, but also those responsible for providing these services are ignorant of the rules and regulations and provide confusing, inconsistent and unfair services, which have important implications for very poor citizens. A further aspect of maladministration is that government officials do not seek to meet their constituents or listen to their concerns. Their lack of interest, in turn, discourages their constituents from interacting with them.

Corruption

It was, however, the strong opinion of the experienced fieldworkers that a failure to deliver the rights and entitlements of citizens was based on attitudes, and these were the result of conscious and proactive corruption. Corrupt government officials sought to augment their income by holding back free services from the poor and turning these into cash for themselves. Teachers and headmasters who invented illegal levies for students and their parents to pay were pocketing the money, as were nurses and doctors who sold free drugs from the government in nearby commercial pharmacies; contractors who paid backhanders to win bids for local construction projects were paying themselves and local government employees extra income taken from government budgets meant for local development.

Few fieldworker participants in social accountability workshops are surprised when the discussions open up into the subject of corruption at local government levels. They know it takes place and they know it is common – in fact, they know that it is the norm. They are frustrated not to know what they can do about it, and they do not believe that the ideas of social accountability will be effective in accomplishing reform. In fact, a number of fieldworkers questioned the value of trying to reverse what they considered as something that was well established and, in practice, inconsequential in the management of local government.

From time to time we had people contribute to the discussion by saying: "Salaries of teachers [or nurses or policemen] are very low and it is to be expected that they need to augment them by *kitu kidogo*" – a Swahili term literally meaning "a little thing" and the accepted phrase to mean a bribe or corrupt payment. Others strongly countered this argument by pointing out that the income of farmers and cattlemen, faced with drought and epidemics, is far worse than that of government officials and they should certainly not have to pay extra fees, which are illegal – even if they are only "a little thing".

What can social accountability do in contexts of pervasive corruption?

The sessions on social accountability often declined into pessimism and feelings that the theory and practice of social accountability could not work in situations of endemic corruption. As the training went on, however, participants started to see where elements of social accountability could be effective – and these were not necessarily the ways in which the policies and practices of social accountability had been taught previously.

The following suggestions come from the experiences of 2018 – what to do, how to do it, and with whom.

1. *What is the hot topic?*

To achieve a serious reform or a major change in practice, fieldworkers accepted that there needs to be a substantial number of people who have all prioritised the same issue and who have a concerted and disciplined approach to trying to get a resolution of the issue that is important to them. It was generally recognised that an individual, or a small group of people, is likely to be intimidated by government officials to the extent that they will give up their cause; however, if government officials appreciate that there are hundreds of people who have the same strongly felt perspective, it is more likely that they will listen to them and, where possible, accommodate them.

We are not talking here about a formal demonstration putting hundreds of people on the streets – which, depending on the laws and the behaviour of police in a particular place, may be dangerous and require a different strategy – but simply letting it be known that there are hundreds of people who feel strongly about a particular topic: for example, teacher absenteeism. The fieldworkers and CSO staff need to do their mobilisation work among those potentially sympathetic to a particular cause, and this is then likely to reach the eyes and ears of government officials. It is important that there is cohesion and consensus about the topic chosen for social accountability action: a community divided between different issues, and possibly split between different party loyalties, is unlikely to achieve its aims.

2. *What information do you need to have?*

This can be separated into two areas: the information you need to have based on the existing Constitution, laws and regulations; and the information you need to collect to prove to government officials what is actually happening.

Very frequently, constitutions, laws and regulations – on the rights of disabled people, for instance – are not easily available, or, where they are available, are not easily understood by local poor and marginalised people: they are in English and in intellectual, academic and legal language. Fieldworkers appreciated the value of making available and disseminating readily understandable, popular versions of important documents on citizens' rights, even to the extent of organising classes to understand them. Such documents, in themselves, can be a spur to action as citizens see the gap between what their rights are and what is actually happening in their lives.

Complementing government regulations and statistics is information on what is actually happening – and this information can be collected by citizens. While the Ministry of Education may say that the maximum number of students in a class is 35, for instance, parents can count attendance and show that the reality is closer to 75. Parents or CSO staff can also easily count the number of teachers in a school at any one time, and how long they are there for – to prove actual teacher attendance figures. This is people's data gathering, sometimes easily done, and at other times likely to provoke a reaction. Citizens can ask to count staff or drugs in a clinic or can ask people exiting a land registration office how much they paid and compare that with the official price. All such information gathered by the people is pertinent when a community decides it is going to try for a reform of the current situation.

3. *Does the government have any understanding of the level of people's satisfaction?*

Government officials who are either newcomers or are long-serving may believe their own government's or department's propaganda and be ignorant about what is actually going on, as well as about people's level of satisfaction with the services they receive. This is particularly true when citizens feel that complaining is a potentially dangerous act. They may, if they complain, be prevented from getting even the poor services about which they are complaining. A regular complaint, for instance, that often came up in conversation was about the rude and inconsiderate behaviour of nurses in clinics. It would take a brave person to complain about this directly to a nurse since it might result in that nurse refusing to serve them thereafter. When a context is set up so that nurses appreciate that many people feel like this and that they all want to improve the situation, then there is the possibility of a more constructive conversation.

In all the situations mentioned above, there is the inherent assumption on the part of citizens and the CSOs that work with them that attitudes and practices which harm the citizens can be opposed, and, if managed judiciously, can be reformed or abolished. The problem is not just learning how to do this effectively,[6] but knowing how to direct your concerns, and in particular to whom.

One regular problem mentioned was never finding the government official that you needed to meet in his or her office; this is important when citizens have travelled long distances to the administrative capital. And, as a result, the person never has the opportunity to have what it is hoped will be a constructive conversation.

A further problem is that citizens' complaints are met with a brick wall: government officials refuse to deliver important government information, simply ignore or rebut citizens' complaints, or threaten those complaining with some form of sanction. For all the well-thought-out techniques and strategies of social accountability, the fieldworkers felt that a constructive conversation was unlikely with those whose income was supplemented by corruption fuelled by the denial of citizens' rights and entitlements. Citizens had to think of strategies which involved complaining to the government officials' supervisors, using media to shame individual government officials, or some form of public denunciation – all of which had potential danger. Sometimes, especially with the police, it was suggested that complaining up the chain of command was useless since the supervisors were taking a cut of the bribes exacted by lower-level officials. They, therefore, had an interest in the system continuing without interruption.

Ethics and integrity with reference to corruption

When these topics were raised, there was great frustration among the fieldworkers I was teaching. While they were enthusiastic that citizens in general and government officials in particular took an ethical stance about citizens' rights and entitlements, particularly those which dealt with the situation of the poor and marginalised, they realised that this was unlikely to translate into action. They referred to the pervasiveness of corruption in society at large, and to the support and approbation that were generally given to those who were successful in their use of corruption. People were considered stupid if they did not take advantage of the opportunities offered for corruption (this point of view was repeated often). Successful people were known for their corruption, and where it was pointed out that corruption caused harm and misery to people, the response was individual charity rather than a radical change to the harmful processes.

Others pointed to the very low salaries that government officials received, so they could not operate without extra income from corrupt practices. However, political activism to improve salaries was not considered as a workable alternative to the continuation of corruption. Many talked of children being introduced to corrupt practices at school when their parents gave them different kinds of bribes to give to their teachers, without commenting that this was wrong and potentially harmful to others.

What did I learn from the coalface?

- That young people are not so enthusiastic about using formal channels that have been laid down and are actually very empowering, if used. They are much more enthusiastic about public protest, even if this results in them being beaten by the police.
- Documentation of failures in public service provision, particularly photos and videos, can be a very powerful tool since these media have an authenticity that personal statements lack.
- Calls for reforms that become national movements, even if they are completely justified, turn many people off – such is the pessimism and distrust of politics in East Africa. Once something becomes popular, like positions for or against dismissing pregnant girls from schools in Tanzania, it becomes politicised – i.e. linked to a political party – which discourages many people from supporting it because of their dislike of the operation of political parties.

What happened to it all?

This chapter is the one that deals with my most recent experiences. I am assured by VSO that practices of social accountability are working, and there are many other active development organisations using these techniques. Unfortunately, however, the readiness of the governments of the three East African countries to answer calls for reform with violence does not seem to change.

And on a personal note

VSO takes a keen institutional interest in the rights of the disabled, and I found myself frequently talking to deaf and disabled people about how they could obtain their rights and entitlements. I will not easily forget the bitterness of groups of deaf people who responded to a senior government official's invitation to talk to them and then found that he had not invited a sign interpreter.

Surprisingly, I was told by disabled people that they were victimised: many ordinary citizens were very ready to take advantage of and to cheat disabled people because it was comparatively easy to do. On the other hand, disabled people were fierce and strongly empowered: "Nothing for us unless done by us," they would chant. This was my first time working with deaf people through sign-language interpreters. The intensity of the communication was astonishing: the straining of deaf people to understand the message and the energy with which they signed back what they wanted to communicate. And, in Kenya, they already had sign language for "Social Accountability" – one sign of fists together to mean "we are doing this together", and one of two fingers on a shoulder to mean "the responsibility is on my shoulders".

What were we thinking at the time?

Research turned up some valuable background information on accountability and local government rules and regulations. There were already some organisations active in the field of social accountability, and I was delighted to find that my videos from Nepal were well received. "This is the same situation that we face," I was told, and further research found valuable local organisations that had done work which resonated with the Nepal experience.

Twaweza. 2017. The untouchables?: Tanzanians' experiences and views of corruption. Brief No. 45. Twaweza East Africa, Dar es Salaam, Tanzania.

Oduor, Chrispine. 2015. *Institutionalising Social Accountability in Devolved Governance*. Institute of Economic Affairs, Nairobi, Kenya.

ISER. 2017. *Are They Effective? An audit of social accountability mechanisms in local government processes in Uganda.* Initiative for Social and Economic Rights (ISER), Kampala, Uganda.

Khadka, Kedar and Bhatterai, Chianjibi. 2013. *Source Book for 21 Social Accountability Tools*. World Bank and PRAN, Kathmandu, Nepal.

See also the PRAN social accountability training videos for Nepal listed in Chapter 17.

Notes

1 The Constitution of Kenya was promulgated in 2010 and a timetable for operationalising devolution developed as Schedule 6 to the Constitution.

True devolution started to take root in 2013 when county government elections were completed and office-bearers elected.

2 Mentioning this induced a rolling of the eyes among the workshop participants as they interpreted this simply as an opportunity to go on government-subsidised shopping trips.

3 "Context Analysis of VSO Kenya's Social Accountability Programme", Institute for Policy and Governance, Kenya, March 2018.

4 I am using demand-side arguments for social accountability (i.e. arguments from the perspective of the citizen). Some development practitioners promote the supply side (i.e. arguments to improve the government's involvement in social accountability). Those with whom I worked in East Africa considered the supply-side approach to be of little value since it was government officials, for the most part, who prevented social accountability from working.

5 Some development workers are surprised that supporters of social accountability do not promote the importance of elections, an accountability mechanism that is common and well established in all three countries. Elections have major shortcomings from the social accountability perspective: they appoint parties and MPs to govern whereas failings in public service provision come from local government officials who are not elected; manifestos – i.e. promises of action if elected – are often buried in personality or ethnic competitions; and people are often paid to vote in a particular way irrespective of manifestos.

6 There are a number of "tools" that those concerned with social accountability have learned to use – such as community scorecards, citizens' report cards, community noticeboards, social or public audits, public expenditure tracking and public hearings with government officials. All manuals on social accountability explain and teach such tools. A problem is that attention to the tools sometimes obscures the fundamental issues of: having community consensus on strongly felt issues; having the community collect data which they are willing to show to others; the community knowing its rights and entitlements; and the community knowing what has been done to prevent its rights and entitlements from being implemented.

20 Reflections

Bringing it all together

... In which I review playing back my experience to the next generation – what kinds of activities still seem to me to have value, and which not: whether I compromised myself; how the aid industry has changed from when I joined it; and how I would like to change it further.

I started this book by saying that it was aimed at actual and would-be development practitioners – fieldworkers and students who are starting out or would like to hear more about the experiences of one long practised in the life they are heading into. I have given examples of what I think has worked, what has not and why, and what is relevant now.

If you have reached this chapter in the book, and perhaps if you have looked at my website (www.richardholloway.org), you will know that I am looking back at my adventures in the aid trade from an advanced age. I have chosen the episodes and the countries because they most appeal to me, and because they illustrate things that I am proud of, things that I consider it was worthwhile to do. There are also references to my wife, also in the aid trade, working for UNICEF, who has supported me, and from whose work I have learned.

The acid test is, I guess, whether my wife and I would encourage our children to go into the aid trade and with what caveats and warnings. There is a short answer: we have done just this and all three children are working with international NGOs (INGOs) in different places and in different ways.

Anything that I might add, however, can only supplement the marvellous book by CDA Collaborative Learning Projects, which listened to 6,000 people involved with the aid trade in 2012 and wrote up what they heard. This book is *Time to Listen: Hearing people on the receiving end of international aid* by Mary Anderson,

Dayna Brown and Isabella Jean. No one involved, or thinking of getting involved, in overseas development work should proceed without a solid reading of that book. I am quoting extensively from it because it strikes me as so genuinely true. The other source of rich thinking is the books and blogs of the chief development guru, Robert Chambers of the Institute of Development Studies (IDS) in Sussex.

More experience in more countries is likely to make you more realistic and perhaps more sceptical.[1] Those of us in the development business tend to talk up our contributions and present ourselves both to ourselves and to our contemporaries as doing worthwhile work, helping people, improving their lives, and helping people make the world a better place. Longer experience shows us how far there is still to go in improving people's lives and how complicated we have made our "industry" as we attempt to do so. On the other hand, many of my children's contemporaries are in the financial services sector and rarely say positive things about their work, or about its contribution to a better world. If pressed and provoked, such people may counterattack with accusations of encouraging autocracy, dependency, corruption and waste, and they may sneer (publicly or not) about do-goodism, blinkered visions or seeking to wear a halo. We have to be able to answer such comments.

It is hard when we are deep in the trenches of aid procedures, when writing or reading endless proposals, budgets, reports, evaluations and audits, to keep our eyes and minds on the sunlit uplands where poor and marginalised people's lives have been improved and where corrupt and despicable people who are trying to interfere with them are beaten back and beaten down. But we must do this in order to energise ourselves, keep sane, and keep a feeling of solidarity with the people whose lives we are trying to improve.

Often, when we are in a capital city of a developing country, it is equally possible to be deep in the trenches of aid procedures there, but we should at least realise that real life is in the street outside – as I discovered in Ethiopia (Chapter 1). This concluding chapter is intended for people who hope that they will be living in the Global South, and what follows are some suggestions for tracks you can make or paths that you might follow that will help you to do things that you are proud of when you are there. Of course, you are not the only actor – none of these things will be possible without committed companions from the country you are working for.

Suggestions of roads to follow

1. Start by being an anthropologist or sociologist

Try to find out what makes the people among whom you are working tick and what aspects of their family and community life could be improved and present opportunities for you to help. One way to do this is by reading histories, anthropologies and contemporary newspapers. Another way is by developing close friendships with locals who will tell you their stories and those of their families. A third is by getting involved in a newish methodology: "Reality Checks".[2] We need to move out of the trap of considering people as "beneficiaries", instead viewing them as equals, peers or clients. VSO likes to use the term "primary actors".

Pursue a desire to learn what makes people tick and try to find out about their indigenous associational life or what they belong to. My guess is that you will be amazed at the depth and importance of associational life for local people. I mentioned in Chapter 9 the women's savings groups in Indonesia called "*arisan*". When conducting savings and credit transactions, the women bring their children, and UNICEF had the smart idea of using this as a vehicle for nutritional education. I mentioned the *Lia Nain* in Timor-Leste (Chapter 14) – the clan-based associations centred on the sacred house which is the first identity of a Timorese. In Tanzania, there are the age groups of the Maasai – the people who have gone through puberty initiation rites together and have bonds all their life. In Mali, there is membership in the Association des Chasseurs, which gives men soulmates for life. Women market traders in Mali have the same kind of relationship. A very valuable exercise is to ask people in the country where you work what they belong to. A broad range of associations will likely be revealed and you will discover why one person defers to another because, for instance, he was in a lower form at school! There will also be the complex world of religious bodies and congregations that people belong to for their spiritual life.

Linked to this is learning how to move your mindset from thinking of poor and marginalised people in the country as *beneficiaries* or "vulnerable" or even "victims" – i.e. the passive recipients of your assistance – and appreciating them as people with capacities and competences which it is your job, working in a development agency, to further develop. The methodology here is called "capacity/vulnerability analysis"[3] and was developed to help fieldworkers dealing with natural (or man-made) disasters, when people were likely to think of themselves, or be thought of by others, as "victims". The methodology encourages you to think about people's competences and capacities to help themselves in three spheres: physical/material, social/organisational and motivational/attitudinal.[4] It is a salutary exercise.

2. *Language*

Overall, and however much we may pay lip service (but often little more than lip service), it is important to learn the language. To learn a language is to learn a culture. Out of all the countries in which I have worked I have learned only one language properly – Indonesian. This is a shameful and embarrassing admission, which is, of course, linked to the amount of time spent in the country. Learning the language will unlock the culture and make you so much more useful. You will not learn the language during short-term consultancies, but if you are there for some time, you should certainly put in the effort to do so. Interestingly, in old colonial days, District Officers were given a pay rise if they could prove that they had learned the local language. Sadly, I have never come across such an idea in any development agency these days.

A lesser achievement, but still one worth trying, is to learn, from your local colleagues, key words or phrases that illustrate concepts that are important to your work and give you insights into people's thinking. I got a lot of mileage in East Africa from learning the Swahili word "*mwana-harakati*". Its positive connotation is "activist" and it is a title that many CSO staff are proud of; its negative connotation is "troublemaker", so it is used by many government officials about CSO people. I already mentioned the unfortunate word "*Morena*" in Chapter 2 on Botswana, and the racism inherent in its use.

If you do not learn the language, the next best thing is to learn to work effectively with interpreters. This is not a small matter: a symbiotic relationship with an interpreter can be extremely useful. In the days when I travelled often to Canada for CUSO, I would arrive jet-lagged and look, befuddled, at night-time TV. Sometimes I saw TV evangelists operating in Latin America with translators, and they were tremendous. You may not like their message, but the choreographed delivery through translators was a lesson to learn from.

3. *Understand people's livelihoods*

In previous chapters I have talked fondly about appropriate technology, but this often leads us to organisations that have designed technologies. I urge exploration of existing indigenous technologies and the trade associations built around particular artisanal groups. There are blacksmiths, potters, weavers and beer brewers who have worked out what technology fits their resources and how to develop it. There are also arable farmers whose indigenous technology involves complicated technical and social arrangements with others to bring irrigation water to their fields, as can be seen in the *subak* of Bali. And at the extreme end there are Thais who build their own Etan[5] trucks.

An understanding of people's use of indigenous technologies gives you an insight into their creativity, collaboration and collectivity. It can also give you an insight into credit and debt. Much of people's livelihoods involve relationships connected with debt and borrowing. These relationships vary over the year as harvests come in, school fees are due, or creditors demand repayment. In some places there is a further associational life connected to the people from whom money can be borrowed, who may be relatives, co-religionists, clan mates or people linked in a variety of other relationships.

4. *Understand what has been done in the name of aid to date*

It is likely that the people with whom you are working will have been exposed to some elements of the aid trade already: this may be government assistance derived from aid to the national government; it may be assistance to citizens that has come from local CSOs supported by INGOs. The listening project *Time to Listen* suggests that while most people are enthusiastic about assistance when it is a new phenomenon, they have negative feelings about the long-term effects of aid and assistance:

> Everywhere people described markedly similar experiences with the processes of assistance and explained how these processes undermined the very goals of the assistance.
>
> 1. Although people want economic support, they see that aid increases dependency and powerlessness.
> 2. Although people want improved security and political stability, they see that aid can worsen conflict and increase tension amongst groups.
> 3. Although people want solidarity and comradeship, they more often feel frustrated, mistrusted, and disrespected by the way aid is provided.[6]

If these findings from the *Time to Listen* study are right, it is quite possible that the people with whom you hope to be working will not be very enthusiastic about more initiatives that come to them in the name of aid.

Past recipients of aid have been recorded as feeling that outsiders are making decisions about them and for them, rather than – and this is the important point – *with them*. Further observations show that there is great frustration with what is seen as the waste of the immense resources available to the aid trade which are meant to help improve their lives.

These longer-term negative feelings about aid are damaging, but they are not necessary or inevitable. If international assistance were to reform itself so that these negatives were no longer felt, then, the listeners claim, aid would become both more helpful and less wasteful.

Thus, it is possible that those offering what they consider to be valuable and positive development assistance may find their interventions to be less than attractive. If this is the case, it will be very important to find out what have been the *un*attractive elements of other people's previous experience with aid, and to try to do something different and better.

5. Finally, understand how damaging corruption has been to date

Corruption is a huge issue – not just the damage that is done to individual families when they must pay to get services which they should be getting for free, but the odium that occurs when it is seen as the regular corollary of aid. On the one hand, local apologists will say that corruption (bribery, nepotism, patronage) is a part of their culture and unlikely to be eradicated; on the other, there are those who consider that aid has actually contributed to corruption.

Those who spoke to *Time to Listen* see corruption as encouraging theft and diversion because there is too much aid and it is distributed too fast. They see that those who own the resources seem loath to follow up on reported corruption and are unenthusiastic about insisting on effective accountability. Even worse, they see that corruption actually translates into international tolerance and complicity.

One aspect of corruption they see is waste – an accusation that is laid at the door of the donors. They say that aid agency people are paid too much in salaries and per diems and that their lifestyle is too wasteful when compared with the proclaimed purpose of aid – to help poor people.

Most donors pride themselves on their due diligence measures, but the feedback coming from *Time to Listen* is that many people are able to circumvent these measures and the donors are not as active as they should be in following up on this. Many remark that the due diligence exercises are set up to prevent corruption, but that if corruption does take place, donors do not look very hard to spot it and do not work very hard to reduce or punish it.

This links to a complaint from local people who manage aid money that they find it very difficult to refuse the pressures on them to share the money with their relatives. In many cultures, better-off members of the clan or family are expected to help less affluent members, and this produces pressure for corrupt measures. They say that if the donors were stricter with their due diligence measures, this would be very helpful for them – they would be supported in their refusals, saying that they can do nothing else. The local project manager could then pass on the strictness, claiming that he or she has no room for manoeuvre.

And where do you stand in all this?

You know that good development aid should involve listening, partnership, collaboration, mutual planning, joint implementation, and clear and participatory reporting, both narrative and financial, and that the whole project or programme should not be so swamped in documenting everything that you cannot do what you know to be good practice – you know all this.

What, then, do you do to change things if you find yourself deep in the trenches and unable to see, let alone visit, the sunlit uplands?

First of all, document yourself and be honest with yourself.

I have hired myself out as a consultant at £500 a day. Others have demanded more. This is not unusual at senior level – it's the market rate – but I have to say that it shames me to be earning that when the people for whom I am working have so little income.

I have been fortunate that I have had employers that I respected – except two (see Chapters 13 and 18). One of these two was a long-term contract that I left, and one a short consultancy that I disliked. I have also been hugely fortunate that the employers that I respected were also ready to pay school fees for my children and so there was no ethical dilemma there. But you have to ask yourself whether you are prepared to stand by your principles when you are part of an organisation that is not doing development well. Either you leave, or you try to find an alternative, or you spend considerable time reforming things from within.

Are you going to be prepared to fight for the right to go on field visits? This seems a strange thing to say, but there is huge pressure in aid organisations to stay in the capital city office and complete the great quantity of paperwork that almost defines the aid industry at present. Paperwork has become more and more laborious and staff-intensive. Field visits are not just for you but are essential for social support to field staff and for monitoring progress – or the lack of it. Will you also demand time and resources to learn the language?

Are you still going to be upright and ethical when the excessive pressures of monitoring and evaluation become unworkable and there is strong pressure to "massage" facts and gloss over realities? What happens when you appreciate that the way you have to deliver aid has got in the way of the purpose of the aid?

How has the aid industry changed in my time?

When I worked for Oxfam in Indonesia in 1979–84, we were told that Oxfam would not take more than 5% from any government; this was so that, if it disagreed with that government's work, it would be able to withdraw from its agreement without too much of a problem. I fully supported that. When I went to work for CRS (Catholic Relief Services) in East Timor in 2002 and met the Oxfam people there, I was astonished to discover that the complete Oxfam programme in Timor-Leste was covered by DFID. People gave me a puzzled look when I asked what happened if they disagreed with DFID policies and practices.

The big bilateral donors have become the main donors in the aid trade. The back donors of the World Bank and the United Nations agencies are the same big bilaterals. These organisations are shifting such large amounts of money that they have become nervous about the competence of big NGOs to handle such amounts, and so the donors have contracted money managers to supervise the spending of their money.

The managers of these large amounts of money for the bilaterals have become more than just money managers; they are also project managers. And for this, in my opinion, they are often not suited. I do not think I am alone in considering that organisations whose expertise has always been in the commercial field of for-profit enterprises – in auditing and then in management consultancy – do not easily fit with not-for-profit development agencies which work in turn with smaller not-for-profit third-sector organisations. These are organisations like Deloitte, PricewaterhouseCoopers, Ernst &

Young and KPMG. The belief seems to be that improving the efficiency of aid delivery will also improve its effectiveness, which I have great difficulty in believing. The five-year project grant from DFID through the PPA (Programme Partnership Arrangement) for the civil society programme of the Aga Khan Foundation (referred to in Chapter 15) was changed in 2010, when the UK Government changed, to a one-year programme to be administered through a for-profit consulting company (Triple Line). This had a new log frame whose main function, as far as I could see, was to be the policeman for more elaborate "value for money" conditions. There was no hint that our previous work was less than satisfactory, so this was not the reason for the change.

One thing that has not changed over my period of work is the "projectisation" of assistance. Donors, for reasons to do with principles of management and administration, decide that aid will come to NGOs and CSOs in one-, two- or three-year cycles, and, after due evaluation, will stop, thus allowing the donor to go on to support something else. You might expect that a positive evaluation would result in an extension of support for a project, but this rarely happens; it seems to be assumed that there is a "natural" life of a project. Unless significant efforts are made during these short-duration projects, they do not develop indigenous ownership and support.

Very unfortunately, the closing of a project is not accompanied by any plans for sustainability. Time after time I have been told that the original donor expects that the project should look for another donor to keep it going. Nor does the original donor invest in building competence in the recipient organisation for raising its own funds in any way other than by seeking further foreign aid.

Vocabulary

There has been an increasing simplification of the vocabulary involved in the aid trade. Some of the phrases are worrying, principally because they impose a business-oriented mindset on a situation which is very different. People in the aid trade should have elements of altruism that are not germane to business mentality and its terminology.

- **"Selling"**: project or programme ideas to donors: the idea here seems to be that recipient organisations (INGOs or CSOs) need to devise attractive ways to interest donors in their ideas. It is a two-person discussion between seller and buyer – there is no space in this thinking for the aid agency to receive the approval of those who finally receive the aid for decisions about further donor funding.

- "**A needs focus**": of development agencies: this suggests that the development agencies will regularly look for situations of need, with the likelihood that they will open up opportunities of dependency. "For a business, the idea of repeat customers provides a steady revenue base: development agencies should be trying to have their clients escape from dependency."[7]
- "**Providers**": "international assistance agencies can supply unwanted goods or services that they, as providers, determine to be 'needed', without apparent penalty, as long as they can sell the idea to their donors".[8] If a business supplied unwanted goods to its customers, it would go out of business.
- "**Growth**": donors, implementing agencies and local partners have come to equate success with growth as an indicator of effectiveness. For a business, delivering some is good; delivering more is better. For development assistance this is not necessarily true. Then there is the vile term "burn rate" to suggest that the rate at which agreed money is spent is important – irrespective of whether it is being used well.
- "**Monitoring and evaluation**": in spite of considerable rhetoric to the contrary, monitoring and evaluation processes focus on what has been done rather than what has happened because of what has been done. What is reported is what they feel is necessary for renewal of the funding.
- "**Delivery**": according to many, this has worryingly become the be-all and end-all of the aid trade. "Even though they do not want to need aid, [those interviewed by the *Time to Listen* study] tell how, as *aid recipients* they develop skills focused on getting the most aid they can, rather than on developing without assistance."[9]
- "**Project entrepreneurs**": fieldworkers, knowledgeable and skilled at understanding good development principles such as participation and partnership, become project entrepreneurs – experts in writing or good at "gaming" the system by appearing to meet the poverty or other criteria they know will "**qualify**" them for aid.
- "**Incentives**": incentives for aid providers to work themselves out of a job – to support people so that they do not need external help (though not all needs may be met) – are almost completely absent compared with incentives to survive, grow and continue to deliver.

- "**Capacity building**": finally, a term I have been guilty of overusing and misusing for years. It should mean building up the competence of someone to do what it is they want to do, but in so many cases it actually means teaching aid recipients to learn aid delivery procedures – particularly financial reporting procedures.

For the most part, I have managed to avoid projects that identified material needs, which would mean spending my time supplying such material needs. I have worked on projects and programmes that responded to discussions with locals about what was not working for them and how some outside training (mixed with training from insiders wherever possible) could be useful. What I have done is:

- manage training in CSO institution building for those who wanted to become competent CSOs (Bangladesh);
- manage training in advocacy for those who wanted to get laws changed (Zambia and Timor-Leste);
- manage training in fundraising and financial sustainability for those who wanted to escape from donor dependency (many places); and
- manage social accountability practices for those who wanted to get away from corruption (Nepal and East Africa).

In all of these, I can legitimately expect that I have worked myself out of a job – through a combination of training and the provision of manuals and learning materials. The project was to place me to do these things; there will be little further expense for those who I have taught to teach others and to implement what they have learned.

And the next generation

- Find good, effective, respected organisations to work for which seem to you to be sincere in their development thinking and which are prepared to push hard for their thinking to be implemented.
- Be certain that any plans to be enacted have been produced with the collaboration of those on whom they are going to be enacted. This is the mantra of people with disabilities – "Nothing done *to* us or *for* us *without* us."[10]
- Be aware of an exit strategy for anything that you enter – how will you leave people better off than when you met them and able to sustain any gains they may have made?

- Research the reality of corruption and see where it is harming poor and marginalised people. I don't know what you will be able to do about it, but move beyond complacency and denial.
- Save money so that you can hold on between jobs and choose only worthwhile ones.

It is an interesting life, and I wish you well.

If you want to learn more

Anderson, Mary B. and Woodrow, Peter J. 1988. *Capacities and Vulnerabilities Workshops: Trainers manual.* Harvard Institute for International Development (HIID), Harvard University, Cambridge MA, USA.

Anderson, Mary B., Brown, Dayna and Jean, Isabella. 2012. *Time to Listen: Hearing people on the receiving end of international aid.* CDA Collaborative Learning Project, Cambridge MA, USA.

Chambers, Robert. 2012. Participation for development: why is this a good time to be alive? IDS blog.

Chambers, Robert. 2014. Perverse payments by results: frogs in a pot and straitjackets for obstacle courses. IDS blog.

Sida. 2012. *Reality Check Bangladesh 2011: Listening to poor people's realities about primary healthcare and primary education – year 5.* Sida, Dhaka, Bangladesh.

Notes

1 Although the always optimistic Robert Chambers said in 2012 that this is a good time to be alive: there are so many wonderful realities and potential for making a difference for the better (see "Participation for development" blog).

2 See *Reality Check Bangladesh 2011.* This documents yearly serial homestays with very poor people over a period of five years to become acquainted with health and education issues from their perspective.

3 Mary Anderson, also one of the authors of *Time to Listen*, co-authored the book on this methodology.

4 Anderson, Mary and Woodrow, Peter. 1988. *Rising from the Ashes: Development strategies in times of disaster.* Harvard Institute for International Development (HIID), Harvard University, Cambridge MA, USA. This was followed by *Disaster and Development Workshops: Manual for training in capacities and vulnerabilities analysis* – by the same authors, but now sadly out of print.

5 Etan trucks were first built in Isarn from recycled material around stationery Yamaha engines. They can carry farm produce – and their engines can be moved and connected to other machinery (pumps, threshers, sawmills) when needed.

6 *Time to Listen: Hearing people on the receiving end of international aid.*
7 Ibid., p. 38.
8 Ibid.
9 Ibid.
10 This is a strong argument against the log frame (logical framework analysis) as it is currently used. When its use was first pioneered, it was constructed with the people who would be affected by the project. However, it is much more common these days for log frames to be designed by the donor without local input.

Index